THE ULTIMATE GUIDE TO REBUILDING A CIVILIZATION

A Technical Manual for the Post-Collapse Era

JAMES EVANS

The Ultimate Guide to Rebuilding a Civilization

LEGAL DISCLAIMER AND SAFETY WARNING

This book contains detailed instructions for chemical synthesis, heavy machinery operation, and medical procedures. It is intended as a repository of knowledge for extreme survival scenarios where modern infrastructure is absent.

Risk of Injury: The processes described involve high temperatures, caustic reagents (sulfuric acid, lye), high-pressure systems (steam engines), and pathogens. Improper execution can result in severe injury, blindness, or death.

Liability: The author and publisher assume no liability for any damages, injuries, or losses incurred during the construction or operation of any device or process described herein. The reader assumes full responsibility for their safety and the safety of their community.

First Edition: January 2026

TABLE OF CONTENTS

I. AGRICULTURE & FOOD

II. MATERIALS

III. METALLURGY

IV. CHEMISTRY

V. POWER & MECHANICS

VI. SCIENCE & SOCIETY

Total 200 Pages of Essential Knowledge

THE BLUEPRINT FOR RESTART

The Knowledge Gap

Civilization relies on specialized supply chains, energy grids, and digital information. System failure locks this knowledge in inaccessible servers or the minds of specialists. This book bridges that gap.

It serves as a repository of foundational technologies. These solutions allow a society to transition from survival to industry. They represent the most efficient, durable, and repairable technologies available when electricity and oil are scarce.

This manual prioritizes function. Each discovery is presented as a standalone unit containing the scientific principle, practical construction instructions, and technical diagrams to guide assembly and operation.

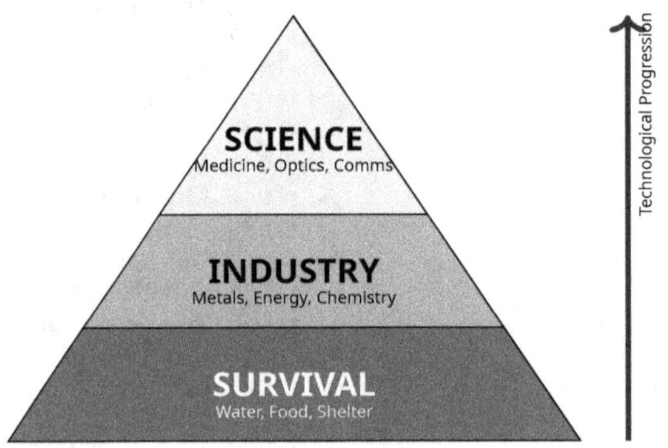

Figure 1: The hierarchy of needs dictates the order of reconstruction.

CORE DISCIPLINES

The Path of Progress

Rebuilding is a sequential process. Building a steam engine requires steel. Making steel requires charcoal. Harvesting enough wood for charcoal requires heavy agriculture.

This guide organizes technologies into logical domains. Mastering one domain provides the tools and materials necessary for the next.

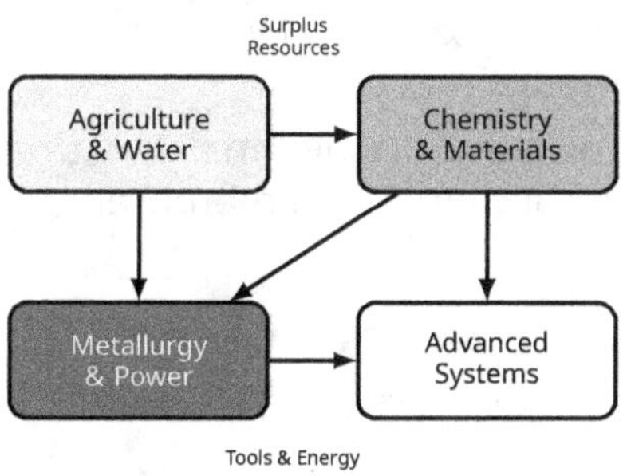

Figure 2: Interdependency of technological domains.

Safety and Responsibility

Many processes described involve high heat, caustic chemicals, and high-pressure systems.

- **Chemistry:** Sulfuric acid and lye cause severe burns.
- **Machinery:** Flywheels and gears store lethal kinetic energy.
- **Biology:** Pathogens are invisible. Assume contamination until proven otherwise.

Knowledge is power. It requires discipline. Proceed with caution, measure twice, and prioritize the safety of your community. The goal is rebuilding.

"The only thing that separates us from the Stone Age is the accumulation of knowledge."

DISCOVERY #001: WATER PURIFICATION

Category: Health / Agriculture **Status:** Essential

The Invisible Killer

In a post-collapse scenario, the greatest threat is not starvation or violence, but waterborne disease. Cholera, Typhoid, and Dysentery kill by dehydration within hours. Clear water is not necessarily clean water; it may be teeming with microorganisms invisible to the naked eye.

Filtration vs. Sterilization

To make water safe, you must address three categories of contaminants:

1. **Sediment:** Dirt and debris (removable by cloth/settling).
2. **Pathogens:** Bacteria, Protozoa, and Viruses.
3. **Chemicals:** Heavy metals and toxins.

A simple cloth filters down to 100 microns. Most bacteria are 0.5 to 5 microns. To stop them without high-tech membranes, we use the **Slow Sand Filter**.

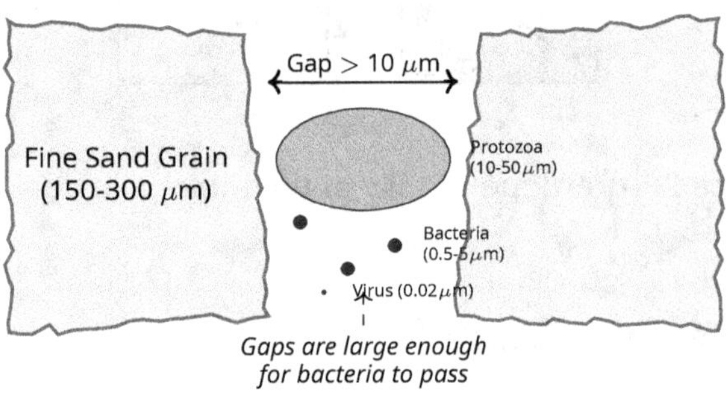

Figure 1: The Scale Problem: Why sieving isn't enough.

The Schmutzdecke: A Living Shield

If sand grains are too large to mechanically catch bacteria, how does a sand filter work? The secret is the *Schmutzdecke* (German for "dirty layer").

When you pour river water through sand, a gelatinous film of bacteria, fungi, and plankton grows on the top 2cm of the sand surface. This biological layer is predatory. The microorganisms in the Schmutzdecke eat the pathogens passing through. They also trap particles via electric charge (adsorption).

Crucial Rule: A new sand filter does NOT work immediately. It requires a "ripening period" of 2-3 weeks to grow this biological layer.

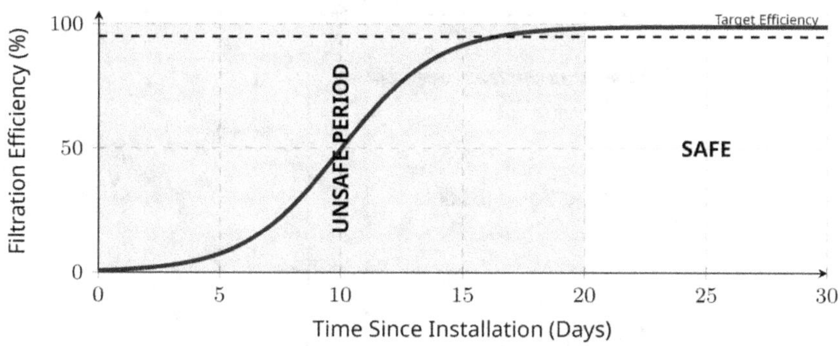

Figure 2: Filter Maturation Cycle. Do not drink water before Day 20.

Operating Conditions

- **Flow Rate:** Must be slow. Max 0.2 liters per hour per cm^2. If it flows too fast, the biological layer cannot trap the bugs.
- **Continuous Wetness:** The biological layer must NEVER dry out. The water level must always stay above the sand.

Construction: The Bio-Sand Filter

You can build this in a plastic barrel, a concrete tank, or a large clay pot.

Materials Needed:

1. **Container:** Approx 1 meter tall.
2. **Diffuser Plate:** A perforated lid or plate to prevent pouring water from disturbing the sand surface.
3. **Fine Sand:** 0.15-0.35 mm grain size. This is the main filter (60cm deep).
4. **Coarse Sand:** Separation layer (5cm).
5. **Gravel:** Drainage layer at the bottom (5cm).

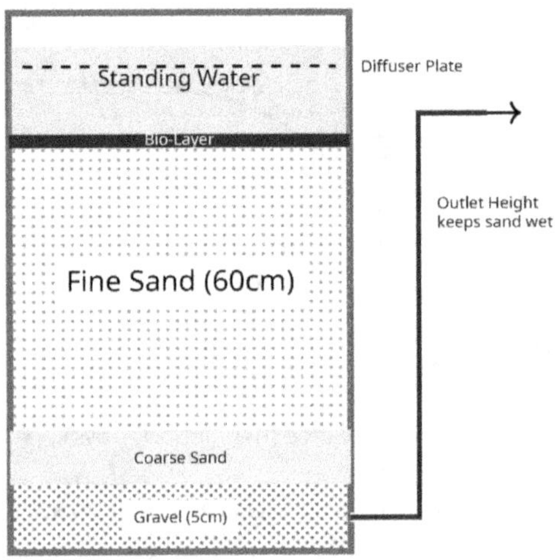

Figure 3: Cross-section of a Household Bio-Sand Filter.

The outlet pipe must rise to a level 5cm above the sand surface. This ensures that gravity stops pulling water when the level hits the pipe height, preventing the sand from ever drying out completely.

Phase 2: Chemical Purification

The sand filter removes 99% of biological threats, but it does not remove chemical toxins or bad tastes. For this, we add a final stage: **Activated Charcoal**.

As we will see in Discovery #004, charcoal has massive surface area. It removes chlorine, pesticides, and organic solvents. Place a cloth bag of crushed charcoal after the sand filter output.

Maintenance Protocol

Eventually, the biological layer will become too thick, and the water flow will stop. Use the "Swirl and Dump" method:

1. Gently agitate the top 2cm of water/sand.
2. The dirty slime will float into the water.
3. Scoop out the dirty water until flow is restored.

Warning: This reduces efficiency for 2-3 days while the layer regrows.

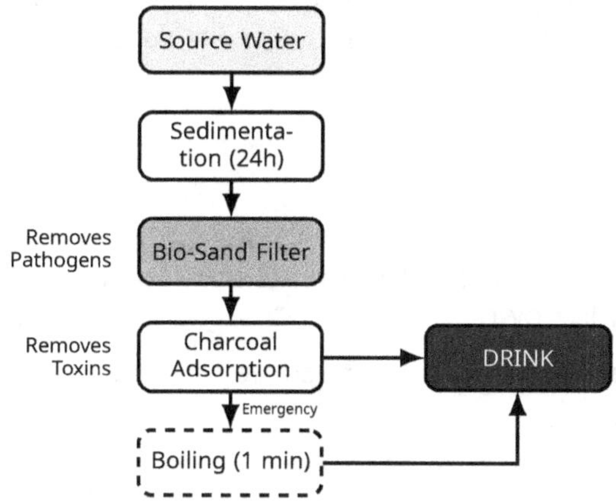

Figure 4: The Complete Water Safety Cycle.

The Golden Rule

If in doubt, boil it. The Bio-Sand filter is for sustainable, daily volume. Boiling is the ultimate guarantee, but it consumes fuel. Combine the two: Filter for general use, boil for the sick or vulnerable.

End of Discovery #001

DISCOVERY #002: THE HEAVY PLOW

Category: Agriculture / Mechanics **Status:** Essential

Breaking the Earth

The simple "scratch plow" (Ard), used by ancient Mediterranean civilizations, is essentially a spiked stick dragged by an ox. It works well on dry, sandy soil. However, if you are rebuilding civilization in temperate, wet climates (like Northern Europe or North America), the soil is heavy, clay-rich, and sticky. An Ard merely scratches the surface, leading to poor drainage and root rot.

To feed a growing population, you need the **Heavy Plow** (*Carruca*). This machine cuts the earth and turns it over completely.

Physics of the Cut

The Heavy Plow performs a complex 3D operation: 1. **Cut Vertically:** Slices the sod. 2. **Cut Horizontally:** Undercuts the root layer. 3. **Inversion:** Flips the soil slice upside down.

This buries weeds (killing them to become fertilizer) and brings fresh nutrients to the surface.

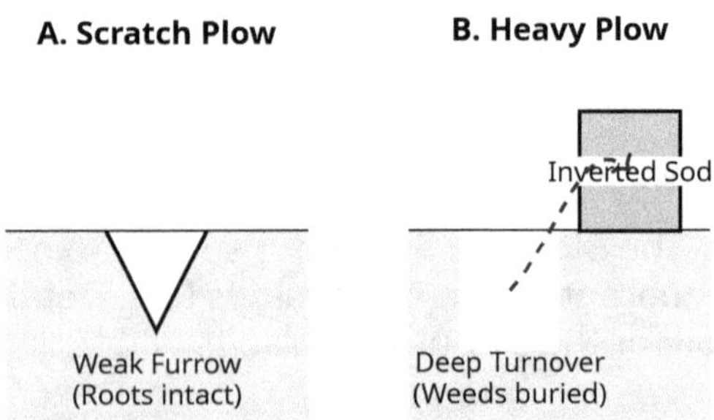

Figure 1: Comparison of soil tillage mechanics.

Anatomy of the Machine

The Heavy Plow is a complex machine requiring precise joinery and iron-working.

The Three Critical Components:

1. **Coulter (The Knife):** A vertical iron blade that cuts the turf. This is the first point of contact.
2. **Share (The Wedge):** A horizontal iron tip that slices under the sod.
3. **Mouldboard (The Hand):** A curved wooden board. This is the innovation. It catches the sliced sod and forces it to roll over to the right side.

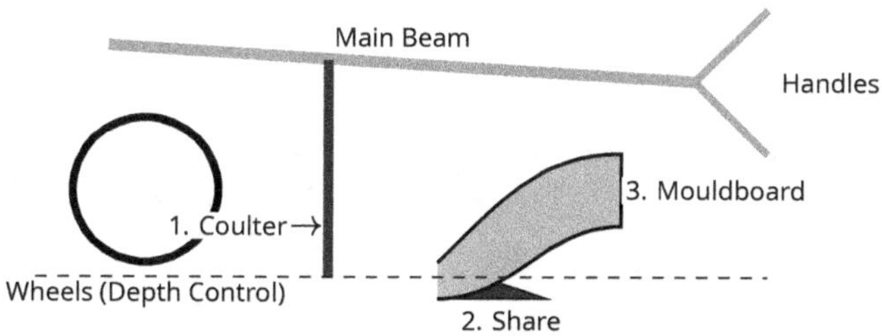

Figure 2: Side elevation of a wheeled heavy plow.

The addition of wheels to the front allows you to set a fixed depth of cut, liberating the plowman from constantly fighting to keep the blade level.

The Ridge and Furrow System

The heavy plow changes the landscape. Because it is heavy and difficult to turn, square fields are inefficient (too many turns). This leads to **Long Fields** (strips).

Furthermore, by always plowing the soil towards the center of the strip year after year, you create a corrugated landscape known as "Ridge and Furrow."

Why this matters: This acts as a built-in water management system.

- **Ridges:** Stay dry in flood years (crops survive).
- **Furrows:** Hold water in drought years (crops survive).

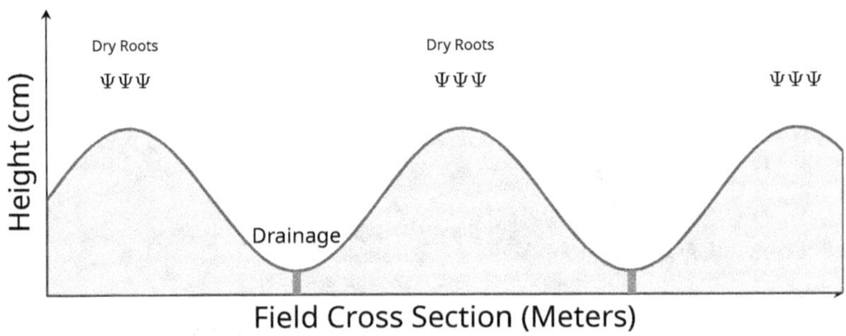

Figure 3: Ridge and Furrow: Hydraulic engineering via plowing.

Community Impact

A heavy plow requires 4 to 8 oxen to pull. No single family owns 8 oxen. This forces the creation of **Communal Farming**. The village must pool resources, leading to shared governance and social stability.

The Power Problem: The Yoke

Having a heavy plow is useless if your animals choke while pulling it. The ancient "Throat-and-Girth" harness used on horses presses on the windpipe. An ox or horse pulling a heavy load with bad harness operates at 20% efficiency.

The Solution: 1. **For Oxen:** The **Withers Yoke**. A wooden beam carved to sit in front of the hump, pushing against the shoulders (skeletal structure), not the neck. 2. **For Horses:** The **Horse Collar**. A padded rigid collar that distributes load to the chest and shoulders.

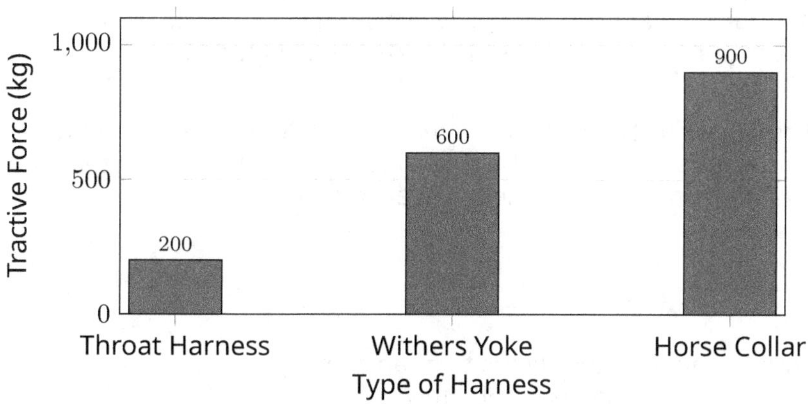

Figure 4: Pulling power available before animal exhaustion.

Summary

The Heavy Plow unlocks the caloric potential of the heavy, wet soils of temperate zones, allowing populations to triple. It creates a landscape resistant to climate variation and enforces social cooperation.

End of Discovery #002

DISCOVERY #003: CROP ROTATION

Category: Agriculture / Biology **Status:** Essential

The Soil Nutrient Limit

Growing cereal crops (wheat, barley, oats) removes nitrogen from the soil. Repeating this process annually depletes the soil's chemical reserves. Historically, farmers addressed this by leaving half their land "fallow" (unused) each year to recover naturally. This practice cuts food production capacity by 50%.

The **Four-Field Rotation System** maintains 100% of land in cultivation every year while simultaneously increasing soil fertility. This system balances nutrient extraction with nutrient replenishment.

Yield Dynamics

Continuous monoculture leads to a rapid collapse in yield. Rotation maintains steady output levels indefinitely.

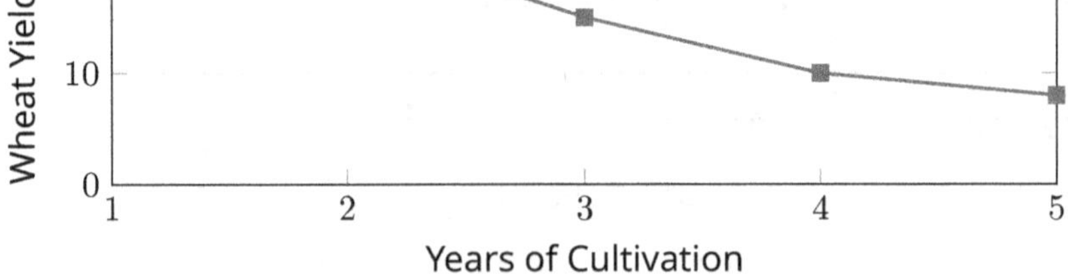

Figure 1: Long-term soil productivity analysis.

The Norfolk Cycle

This specific sequence optimizes pest control, weed suppression, and nutrient management. It repeats every four years on each plot of land.

Year 1: Wheat (Cash Crop) The primary calorie source for humans. It demands high nitrogen and clean soil.

Year 2: Turnips (Fodder/Cleaning) Turnips have deep roots that break up the soil. They are fed to livestock. Importantly, turnip farming requires "hoeing" during growth, which physically removes weeds that established themselves during the wheat year.

Year 3: Barley (Spring Crop) Barley is less demanding than wheat. It provides grain for beer (safe water) and animal feed.

Year 4: Clover (Recovery) Clover is a legume. It is grown to feed livestock and, crucially, to restore nitrogen to the soil for the next wheat cycle.

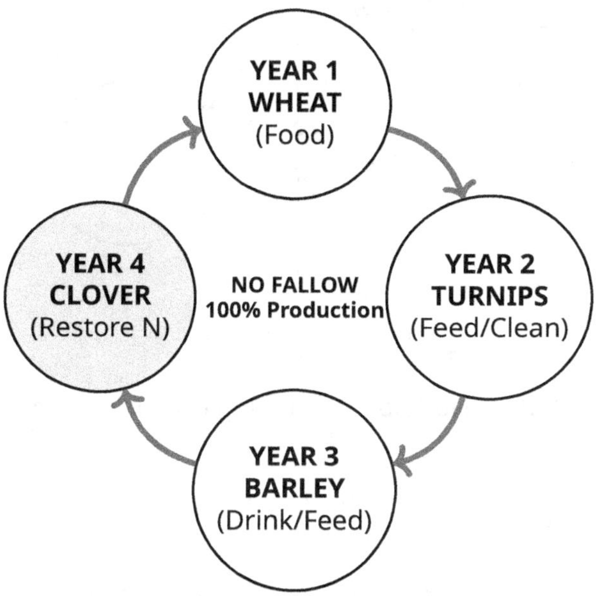

Figure 2: The Norfolk Four-Field System.

Nitrogen Fixation

Clover restores fertility through a symbiotic relationship with bacteria.

Atmospheric air is 78% nitrogen, but plants cannot digest gas. They require nitrates. Legumes (clover, beans, peas) host *Rhizobia* bacteria in their roots. These bacteria absorb atmospheric nitrogen (N_2) and convert it into ammonia (NH_3), which plants can absorb.

When the clover plant dies or is plowed back into the earth ("Green Manure"), this stored nitrogen is released into the soil, ready to fuel the wheat crop in Year 1.

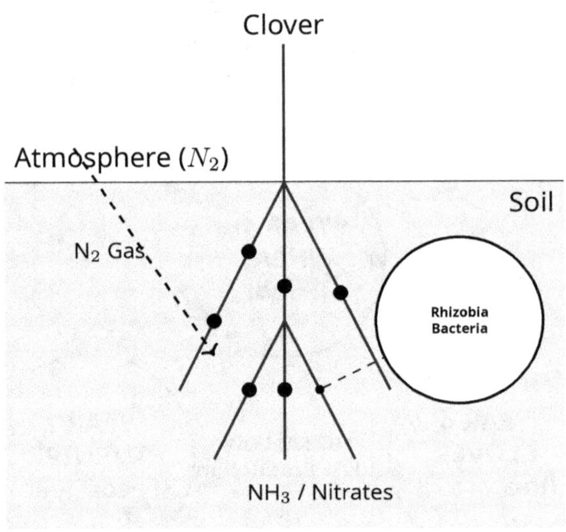

Figure 3: Root Nodules: The biological fertilizer factories.

Green Manure

The nitrogen payload is concentrated in the roots. Even if the foliage is harvested for animal feed (as in the Norfolk system), the nitrogen-rich root mass remains. Plowing this biomass back into the soil creates the perfect seedbed for the subsequent wheat crop.

The Winter Bottleneck

Historically, farmers slaughtered most livestock in November because grass does not grow in winter and they lacked fodder. This resulted in a lack of manure (fertilizer) in spring and a loss of animal capital.

The Solution: Turnips and Clover. These crops are preserved as winter feed.

1. **Turnips:** Store well in root cellars or clamps.
2. **Clover Hay:** Dried nutrient-dense fodder.

This capability allows herds to survive the winter intact, leading to larger herds, more meat, more milk, and significantly more manure.

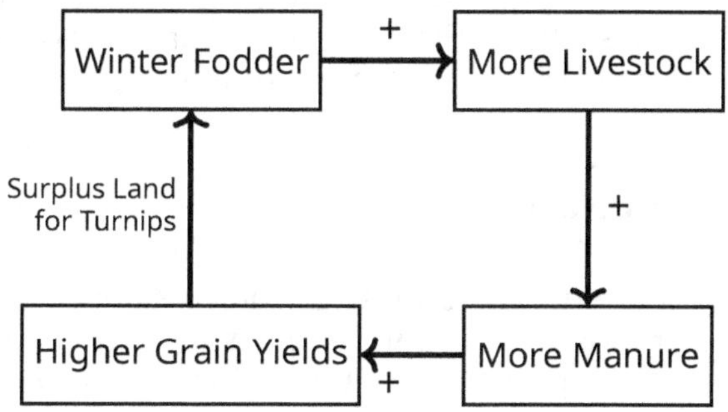

Figure 4: The Virtuous Cycle of Integrated Farming.

Summary

Crop rotation integrates biology (nitrogen fixation) and animal husbandry into a unified machine. It eliminates the need for fallow land and breaks the bottleneck of winter starvation, creating a surplus of both grain and protein.

End of Discovery #003

DISCOVERY #004: FOOD PRESERVATION

Category: Chemistry / Survival **Status:** Essential

The Biological Timer

Fresh meat and vegetables decay because bacteria and fungi consume them. These microorganisms require moisture to survive and reproduce. Removing moisture stops biological activity.

Salting preserves food through **Osmosis**. Salt is hygroscopic; it attracts water molecules. When you cover meat in salt, the salt draws water out of the bacterial cells. This causes the bacteria to shrivel and die (plasmolysis). The food becomes a hostile environment for life.

Mechanism of Action

To effectively preserve meat, you must achieve a specific salinity concentration. A concentration of 10% salt removes enough water (a_w or "water activity") to inhibit most pathogens. 20% inhibits virtually all life.

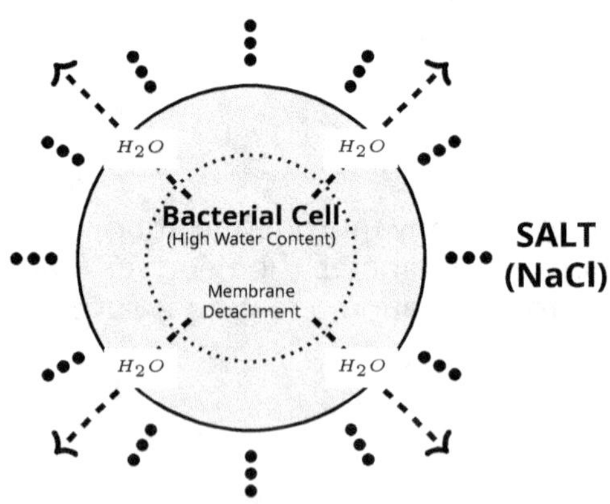

Figure 1: Plasmolysis: Salt extracts cellular water, killing bacteria.

Chemical Preservation: Smoking

Smoking works in tandem with salting. While heat dries the meat, the smoke deposits chemical compounds that act as preservatives. Wood smoke contains hundreds of compounds.

Key Components of Smoke:

1. **Phenols:** Powerful antioxidants that prevent fats from going rancid. They also kill bacteria on the surface.
2. **Formaldehyde & Acetic Acid:** Lower the pH of the meat surface, making it acidic and antimicrobial.
3. **Creosote:** Forms a physical barrier (the "pellicle") that seals the meat against insects.

Cold vs. Hot Smoking

Cold Smoking (<30°C): Preserves raw meat for long-term storage. Requires days or weeks. **Hot Smoking (60-80°C):** Cooks the meat. Good for flavor and short-term extension, but does not dehydrate sufficiently for winter storage.

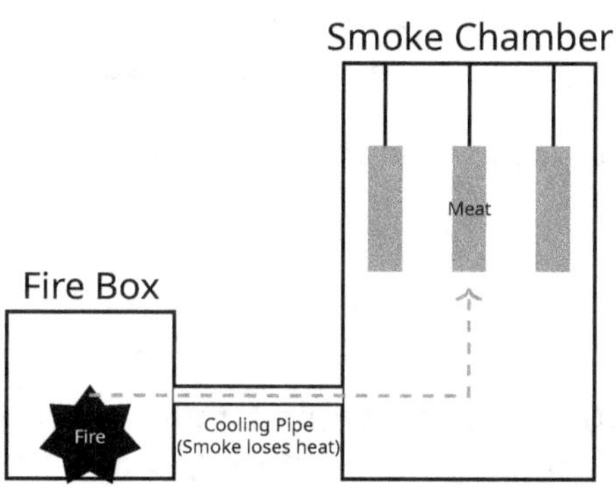

Figure 2: Cold Smoker Design: Smoke reaches food without heat.

Controlled Decay: Fermentation

Salting and smoking kill bacteria. Fermentation *recruits* beneficial bacteria to kill harmful ones.

Lactobacillus bacteria are naturally present on cabbage leaves. In an anaerobic (oxygen-free) environment, they convert carbohydrates (sugars) into **Lactic Acid**.

This process preserves food because Lactic Acid lowers the pH of the food. Clostridium botulinum (which produces botulism toxin) cannot survive in acidic environments (pH < 4.6). Fermented foods like Sauerkraut are safe, shelf-stable, and rich in Vitamin C.

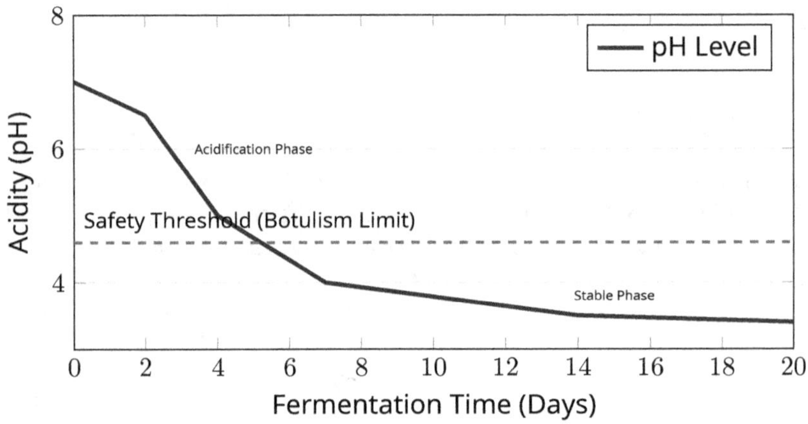

Figure 3: The Acidification Curve: Reaching safety within one week.

Nutritional Value

Sauerkraut retains 100% of the Vitamin C found in fresh cabbage. Historically, sailors used it to prevent Scurvy during long winters or voyages.

The Production Process

To make Sauerkraut, you need cabbage, salt, and a container.

The Ratio: Use 2% to 3% salt by weight (20g salt per 1kg cabbage). **The Method:** 1. Shred cabbage finely to increase surface area. 2. Massage salt into the cabbage until it releases its own water (brine). 3. Pack tightly into a crock. 4. Weight it down. **The solid food must stay submerged under the liquid.**

The Water Seal

Oxygen is the enemy. Oxygen allows mold and aerobic bacteria to grow. You must create an airlock that lets CO_2 escape while preventing fresh air from entering.

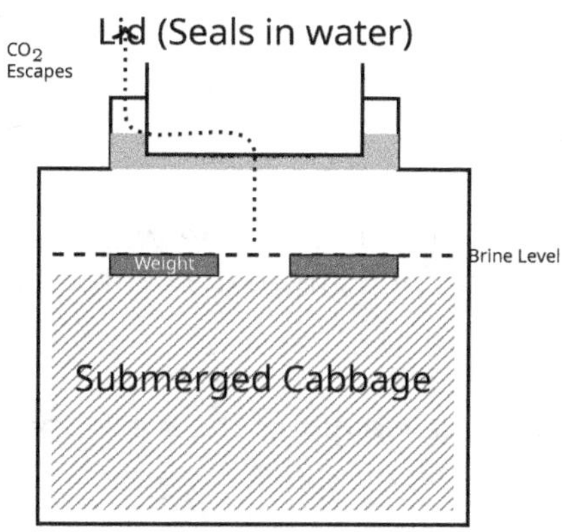

Figure 4: Fermentation Crock with Water Seal.

Summary

Preservation transforms a harvest surplus into a survival stockpile. By controlling moisture (Salting), chemical environment (Smoking), or acidity (Fermentation), you deny pathogens the opportunity to grow.

End of Discovery #004

DISCOVERY #005: THE WATER WHEEL

Category: Energy / Mechanics **Status:** Essential

Free Energy

Grinding grain into flour requires immense physical labor. A person using a hand quern processes approximately 5kg of grain per hour. A water mill processes 150kg per hour or more, freeing the population for other tasks.

The Water Wheel converts the kinetic energy of moving water or the potential energy of falling water into rotational mechanical energy. This torque drives machinery for milling flour, sawing timber, or pumping bellows for iron smelting.

Physics of Rotation

The mechanism relies on Torque (τ). The water exerts force (F) on the blades or buckets at a distance (r) from the center axle.

$$\tau = F \times r$$

Maximizing the radius or the weight of the water increases the power output.

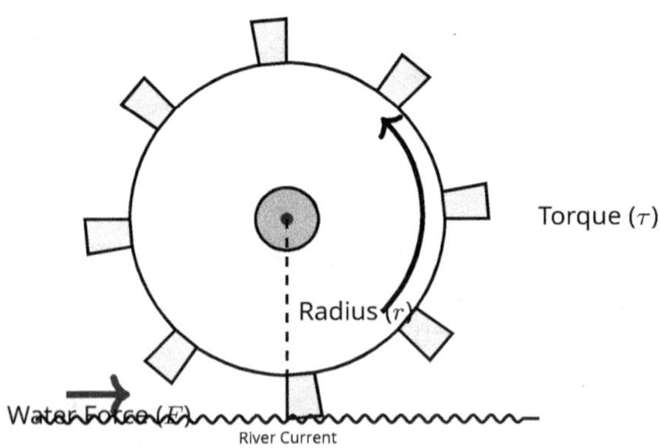

Figure 1: Converting linear flow into rotational torque.

Wheel Configurations

The efficiency of the mill depends on how the water strikes the wheel.

1. The Undershot Wheel The bottom of the wheel dips into a flowing stream. It relies entirely on the speed (kinetic energy) of the river.

- **Efficiency:** Low (30%).
- **Construction:** Simple. No dam required.

2. The Overshot Wheel Water is channeled via a flume to the top of the wheel. The water fills buckets. Gravity pulls the heavy, full buckets down while empty buckets rise. This utilizes the weight (potential energy) of the water.

- **Efficiency:** High (85%).
- **Construction:** Complex. Requires a headrace (channel) and elevation drop.

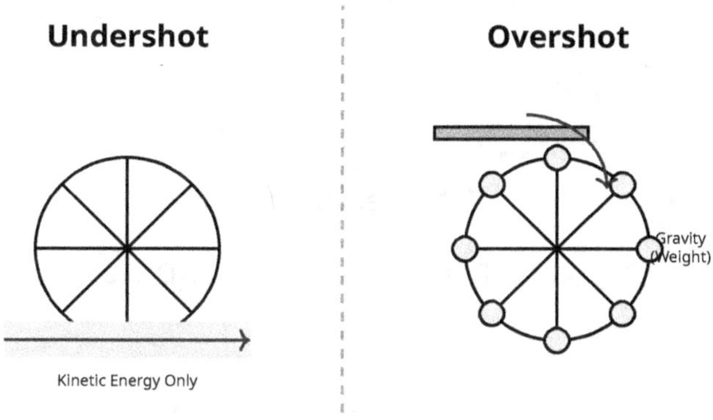

Figure 2: Design Comparison: Impact vs. Gravity.

The Overshot wheel provides superior power output even with low water flow, making it the preferred choice for industrial applications.

Transmission and Gearing

A water wheel rotates slowly (5-10 RPM). A millstone must rotate rapidly (60-120 RPM) to grind effectively. Furthermore, the main axle is horizontal, but the millstone spins on a vertical axis.

The Gear Train solves both problems: 1. **The Pit Wheel:** A large vertical gear attached to the main axle. 2. **The Wallower:** A small horizontal gear (lantern gear) that meshes with the Pit Wheel.

This arrangement multiplies speed (large gear driving small gear) and translates rotation by 90 degrees.

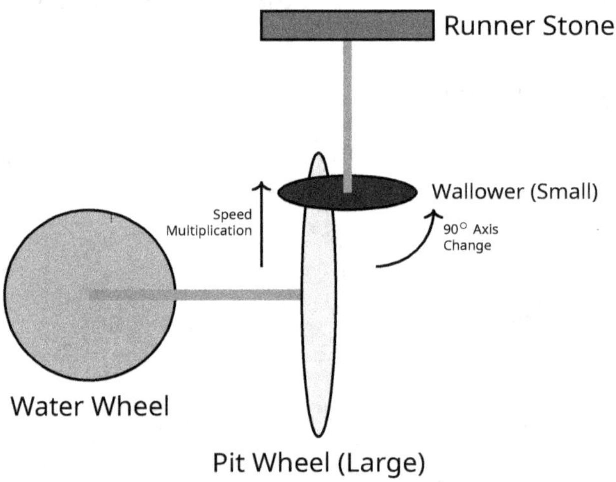

Figure 3: The Right-Angle Gear Drive.

Material Requirements

The gears endure immense stress. The teeth (cogs) are made from the hardest available wood (Hickory, Hornbeam, or Oak) and are replaceable. They act as the "fuse" of the system, breaking before the main axle snaps if the mill jams.

The Grinding Mechanism

The purpose of the entire facility is to turn the **Runner Stone**. 1. **The Bedstone:** Fixed permanently to the floor. 2. **The Runner Stone:** Spins slightly above the Bedstone.

Grain falls from a **Hopper** into the center hole ("eye") of the Runner Stone. Centrifugal force pushes the grain outward between the stones.

The Furrows: The stones are not smooth. They are carved with a pattern of grooves called furrows. These act like scissors to shear the grain husk and channels to push the flour out to the rim.

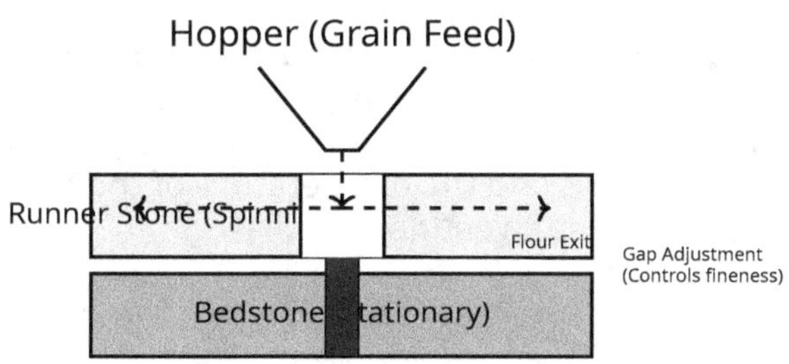

Figure 4: Cross-section of the Stone Floor.

Summary

The water mill represents the transition from muscle power to machine power. It provides the surplus energy foundation required for advanced metallurgy, textile production, and large-scale agriculture.

End of Discovery #005

DISCOVERY #006: PASSIVE REFRIGERATION

Category: Physics / Food | **Status:** Essential

The Thermodynamics of Cool

Bacteria reproduction rates drop significantly below 10°C (50°F). Maintaining this temperature without electricity requires leveraging phase change physics.

The primary mechanism for low-tech cooling is **Evaporative Cooling**. When water transforms from a liquid to a gas, it absorbs thermal energy from its surroundings to break the molecular bonds. This process extracts heat from the environment, lowering the temperature of the remaining surface. This is the same principle as sweating.

Latent Heat of Vaporization

Water has a high latent heat of vaporization (2260 kJ/kg). Evaporating a small amount of water removes a large amount of heat energy from the container holding it.

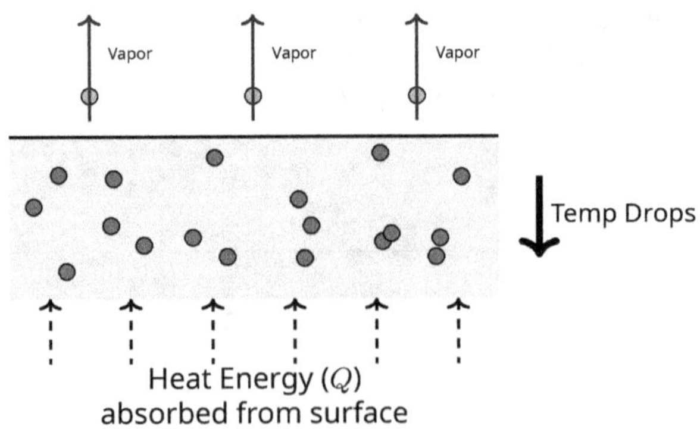

Figure 1: Energy transfer during evaporation.

The Zeer Pot (Pot-in-Pot)

The Zeer Pot utilizes evaporative cooling to preserve vegetables in hot, dry climates. It consists of two earthenware pots, one nested inside the other, with a layer of wet sand in between.

Construction: 1. **Outer Pot:** Porous earthenware (unglazed). This allows water to migrate to the surface. 2. **Inner Pot:** Glazed or sealed interior. This prevents water from contaminating the food. 3. **Medium:** Sand fills the gap. Sand holds water and provides surface area for evaporation.

As water evaporates from the outer surface, it pulls heat from the inner pot. In a dry climate with a breeze, this can cool the inner chamber to 15°C below ambient temperature.

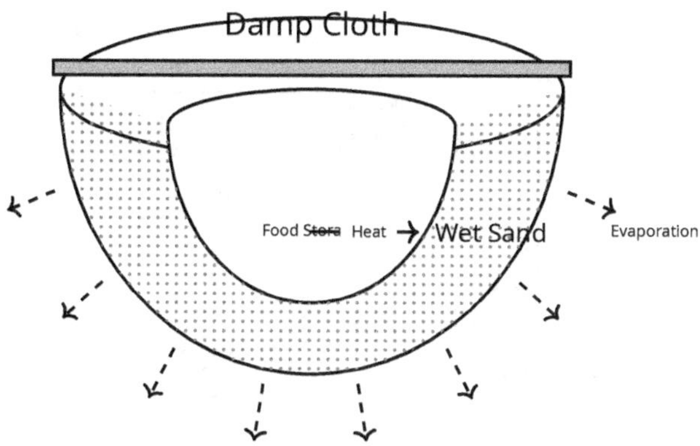

Figure 2: Cross-section of a Zeer Pot refrigerator.

Industrial Cooling: The Ice House

While the Zeer Pot creates cool air, the Ice House *stores* winter cold for summer use. This relies on **Thermal Mass** and massive insulation.

Water must be harvested from frozen lakes or rivers in winter. The ice is cut into large uniform blocks. Large blocks have a lower surface-area-to-volume ratio than small chunks, reducing the melting rate.

Structure: The facility is built underground to utilize the stable ground temperature. The ice is packed tightly in a central pit. The critical design feature is **Drainage**. Standing water conducts heat 25 times faster than air; melted water must be removed immediately to preserve the remaining ice.

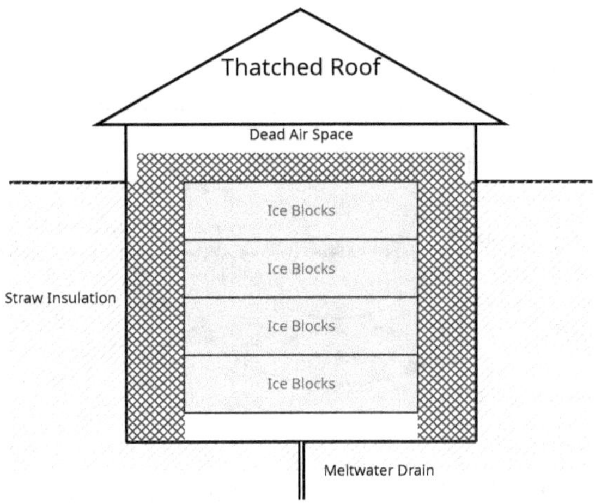

Figure 3: Underground Ice House with drainage and insulation.

Insulation Dynamics

Insulation works by trapping pockets of dead air. Still air is a poor conductor of heat.

Materials:

- **Sawdust:** The most efficient pre-industrial insulator. Tightly packed sawdust prevents air convection.
- **Straw:** Good, but requires a thicker layer (30-50 cm).
- **Wood:** Moderate insulator, provides structure.

With 30cm of sawdust insulation and proper drainage, an ice house loses less than 1% of its mass per day in summer.

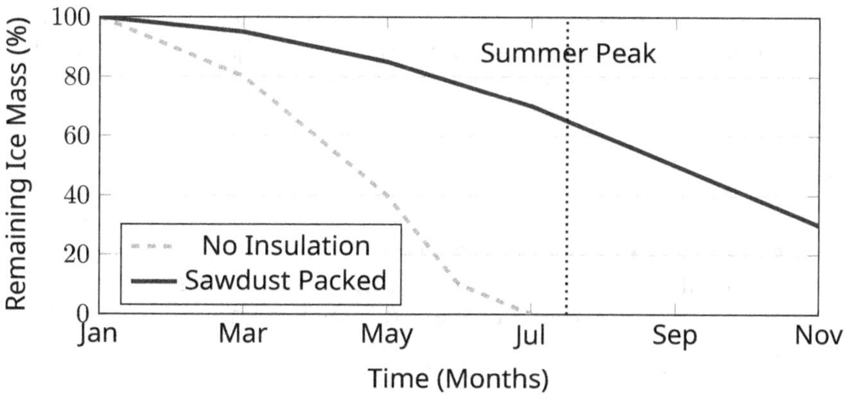

Figure 4: Retention rates based on insulation quality.

Summary

Refrigeration allows for the accumulation of surplus protein (meat/dairy) and the preservation of medicines. Whether through the physics of evaporation or the storage of winter ice, controlling temperature breaks the cycle of bacterial decay.

End of Discovery #006

DISCOVERY #007: ARTIFICIAL SELECTION

Category: Agriculture / Genetics **Status:** Essential

Accelerating Nature

Wild plants evolve to survive, not to feed humans. They produce small, bitter seeds that scatter easily. To rebuild agriculture, you must direct the evolution of your crops.

Artificial selection is the process of breeding only those individuals that possess desirable traits. By saving seeds exclusively from the best 5% of your harvest, you shift the genetic average of the population. Over just five generations, this focused pressure transforms a wild grass into a productive grain or a small root into a calorie-dense vegetable.

The Bell Curve Shift

Most traits (size, sweetness, yield) follow a normal distribution. In every field, a few plants naturally outperform the average due to genetic variation. Saving seeds from these outliers moves the entire curve to the right in the next generation.

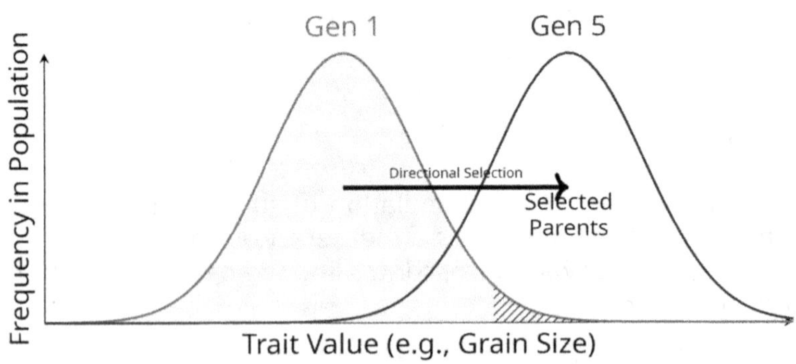

Figure 1: Shifting the genetic baseline through selective breeding.

Domesticated Traits

A good crop plant differs physically from its wild ancestors. You must observe your fields and tag plants that exhibit specific "Domestication Syndrome" traits.

Primary Targets:

- **Non-Shattering Rachis:** In wild wheat, the seed head explodes (shatters) to spread seeds. In domesticated wheat, the seeds stay attached until harvested by hand.
- **Determinate Growth:** The entire field ripens simultaneously, allowing for a single efficient harvest.
- **Gigantism:** Selection for larger fruits or seeds often results in fewer, larger organs.

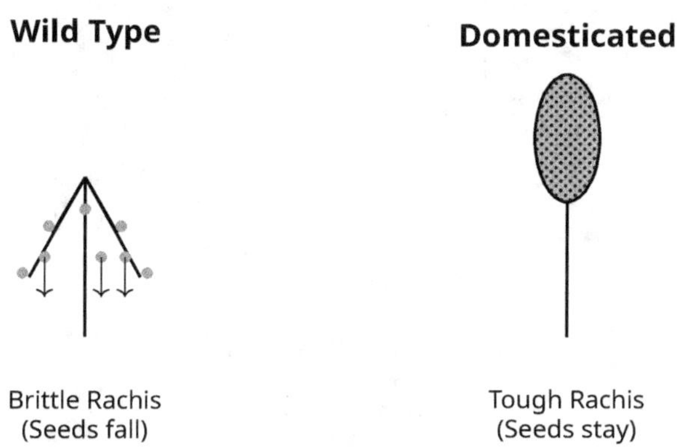

Figure 2: The critical mutation for agriculture: Seed Retention.

Always prioritize health and resistance over simple size. A giant pumpkin that rots easily is useless for survival.

Maintaining Purity

Plants reproduce sexually. Pollen from a wild, bitter variety can fertilize your sweet, domesticated variety, ruining the next generation's seeds.

Pollination Mechanisms:

- **Self-Pollinating (Easy):** Wheat, Beans, Peas, Tomatoes. The flower fertilizes itself before opening. Requires minimal isolation (3-5 meters).
- **Cross-Pollinating (Hard):** Corn, Squash, Cabbage. Pollen travels by wind or insects for kilometers.

To save pure seed from cross-pollinators, you must isolate them by distance or time.

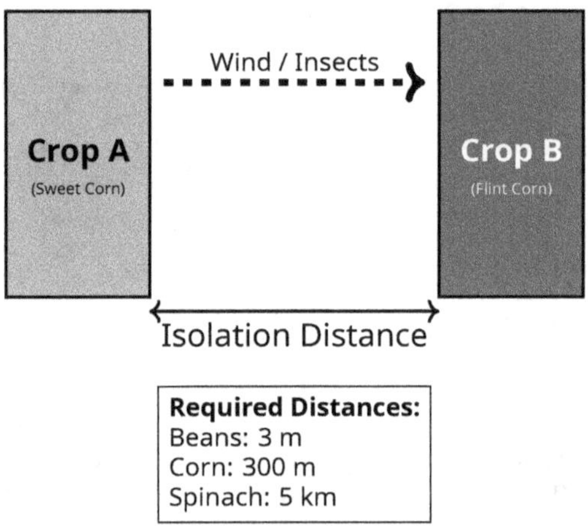

Figure 3: Preventing genetic contamination.

Mechanical Isolation

If distance is impossible, cage the plants with insect netting or bag individual flowers before they open, pollinating them by hand.

The Seed Bank

A seed is a living embryo in suspended animation. It consumes its own energy reserves to stay alive. Heat and moisture speed up this metabolism, causing the seed to burn out and die.

The Rule of 100: The sum of the Storage Temperature (°F) and Relative Humidity (Ideally: **Cold (5°C)** and **Dry (<10% Moisture)**.

The Jar Test: Place seeds in a sealed glass jar with dry salt. If the salt clumps after 24 hours, the seeds are too wet to store. Dry them further in the shade (never direct sun) before freezing.

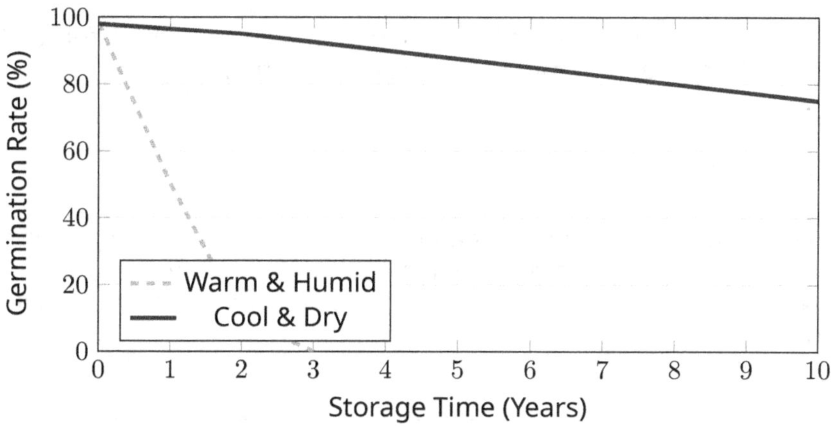

Figure 4: Impact of storage conditions on viability.

Summary

Civilization is built on the genetic achievements of ancestors. Saving the best seeds is the primary duty of the farmer. It ensures that next year's harvest will be more abundant and resilient than the last.

End of Discovery #007

DISCOVERY #008: ADVANCED FIRE

Category: Energy / Metallurgy **Status:** Essential

Beyond the Campfire

An open fire loses 90% of its energy to the atmosphere through convection and radiation. It reaches a maximum temperature of approximately 600-700°C. This is insufficient for smelting copper (1085°C) or iron (1538°C).

To achieve metallurgical temperatures, you must trap heat. A furnace is a thermal battery that accumulates energy faster than it loses it.

The Insulation Principle

Thick walls made of low-density material (mud, clay, straw) slow the transfer of heat from the inside to the outside. As the fire continues to burn, the inner walls absorb energy and re-radiate it back into the center, creating a positive feedback loop that drives temperatures upward.

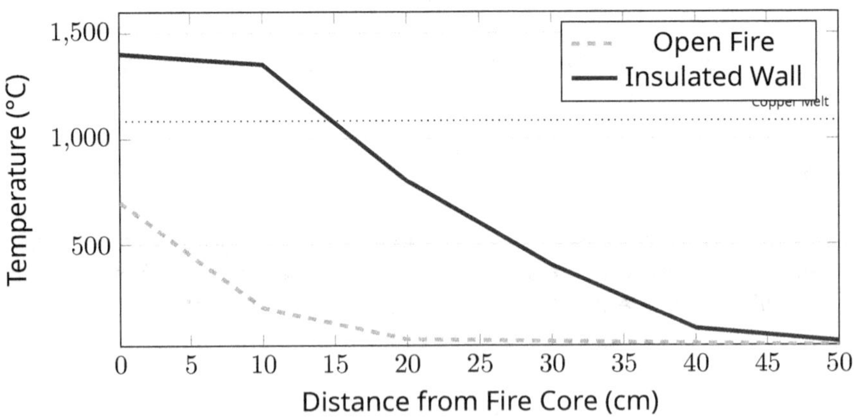

Figure 1: Effect of insulation on core temperature.

The Chimney Effect

Fire requires three things: Fuel, Heat, and Oxygen. In a high-temperature furnace, Oxygen is the limiting factor.

Hot air is less dense than cold air and rises. By building a vertical shaft (chimney) above the fire, you create a low-pressure zone at the bottom. This vacuum pulls fresh, oxygen-rich air into the intake vents at high speed. A taller chimney creates a stronger draft, increasing the rate of combustion without human effort.

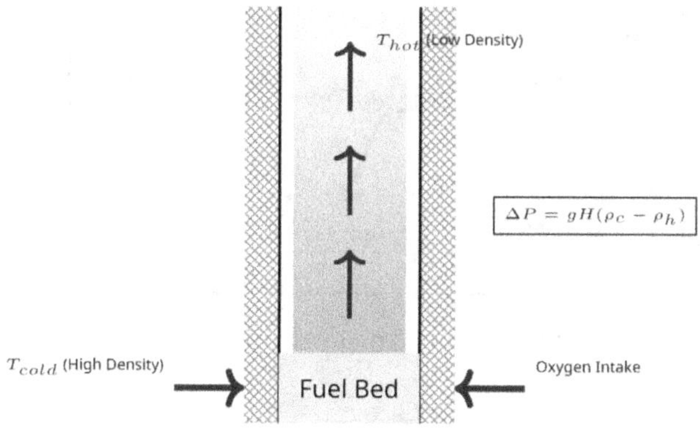

Figure 2: The Chimney Effect: Converting heat height into airflow.

For temperatures above 1300°C, natural draft is insufficient. You must use **Bellows** (see Discovery #029) to force air into the core.

Reflecting Heat

In primitive smelting, fuel mixes with ore. This contaminates the metal with ash and sulfur. Ideally, the fuel should be separate from the workpiece.

The **Reverberatory Furnace** solves this. It has a separate firebox. The heat rises, hits a curved roof (arch), and reflects ("reverberates") down onto the metal hearth. This allows for clean melting of glass, bronze, and iron.

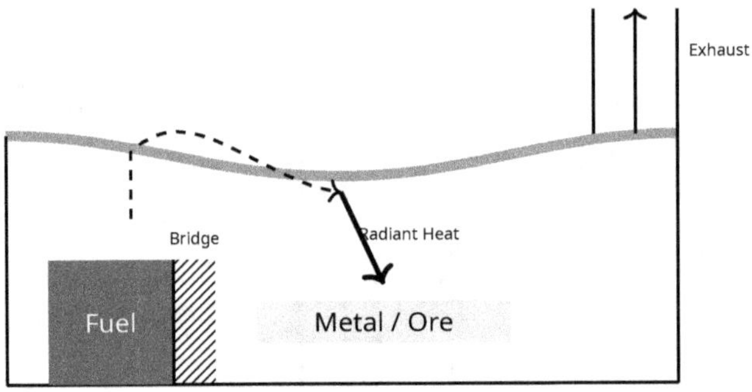

Figure 3: Separating fuel from product via reflected energy.

The Bridge Wall

A low wall separates the fire from the metal. It forces the flame to leap over it, ensuring only hot gases (not solid ash) reach the refining chamber.

Refractory Materials

Ordinary river rocks contain water. When heated rapidly, they explode. Ordinary clay melts at 1000°C, turning your furnace into a puddle of slag.

To build a high-temperature furnace, you need **Refractory Brick** (Firebrick). This requires clay with high Alumina (Al_2O_3) and Silica (SiO_2) content, and very low impurities (lime or iron oxide).

Finding Fireclay: Look for gray or white clays under coal seams. Unlike red clay (rich in iron), fireclay withstands 1600°C.

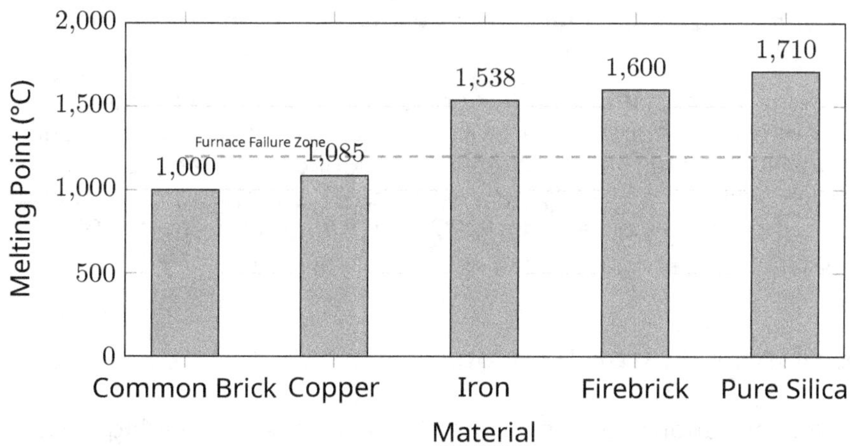

Figure 4: Material limits dictate maximum temperature.

Summary

Fire is a tool, not just a reaction. By controlling insulation, airflow, and materials, you transform simple combustion into an industrial force capable of liquefying rock and forging civilization.

End of Discovery #008

DISCOVERY #009: CHARCOAL

Category: Metallurgy / Materials **Status:** Essential

The Thermal Barrier

Raw wood burns at approximately 700°C. While sufficient for cooking, this temperature fails to melt copper (1085°C) or iron (1538°C). To enter the metal age, civilization requires a fuel with a higher energy density and purity.

Charcoal is wood that has undergone **Pyrolysis**: the decomposition of organic material at high temperatures in the absence of oxygen. This process drives off water, volatile organic compounds, and tars, leaving behind a porous skeleton of nearly pure carbon. This fuel burns clean and can reach temperatures exceeding 1200°C with natural draft, or 2000°C with forced air.

Chemistry of Carbonization

The transformation occurs in distinct thermal stages. Managing these stages ensures a high yield of solid fuel rather than ash.

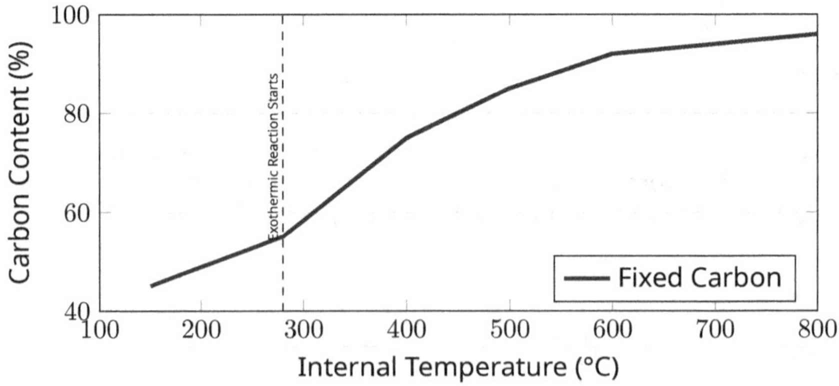

Figure 1: Purity increases as volatiles are driven off.

Wood Selection

The quality of the charcoal mirrors the density of the source wood.

Hardwoods (Oak, Hickory, Beech): These produce dense, heavy charcoal that burns slowly and intensely. This is the standard fuel for blacksmithing and smelting, where structural strength is required to support the weight of ore in a furnace.

Softwoods (Pine, Willow): These produce light, friable charcoal. It ignites instantly but burns fast. While poor for smelting, it is the superior choice for making black powder (gunpowder) due to its high porosity.

Seasoning: Wood must be dried (seasoned) for 6-12 months before charring. Green wood contains up to 50% water. Burning wet wood wastes massive amounts of energy boiling off the water before carbonization begins.

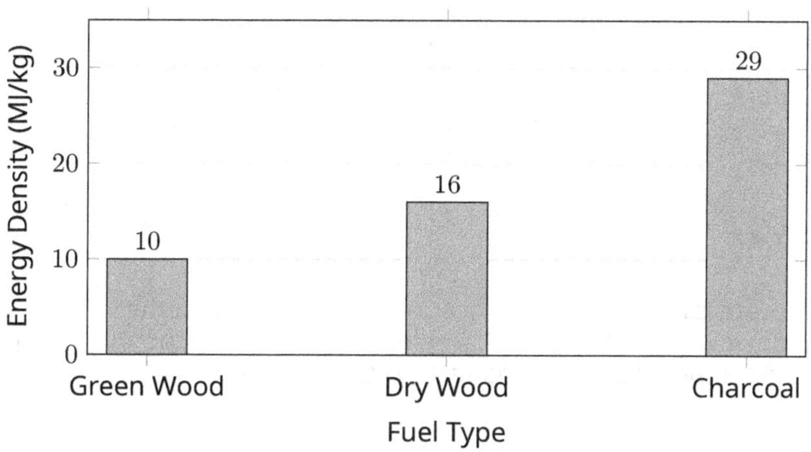

Figure 2: Charcoal nearly doubles the energy per kilogram.

Construction: The Earth Mound

The most accessible method for mass production is the Earth Mound Kiln. It requires no bricks, only labor and careful stacking.

The Structure: 1. **Chimney:** A central shaft provides the initial draft and ignition point. 2. **Stacking:** Wood is stacked vertically to encourage airflow from the base to the center. 3. **Mantle:** A crucial insulation layer composed of straw and wet earth. This prevents oxygen from entering the pile while retaining heat.

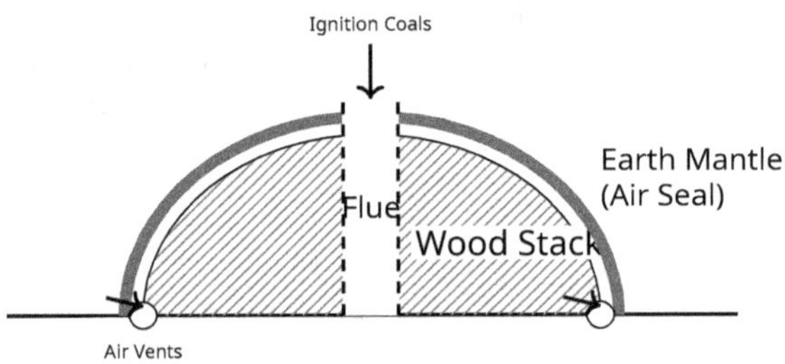

Figure 3: Cross-section of a standard mound kiln.

The Burn

The process is a controlled suffocation. The fire starts at the top center and burns downward and outward. The operator controls the rate of combustion by opening or closing small air vents at the base.

Reading the Smoke

Visual indicators are the only telemetry available. The color of the smoke reveals the chemical stage of the reaction inside the mound.

1. White Smoke: Steam. The wood is drying. Keep vents open. **2. Yellow/Thick Smoke:** Pyrolysis. Tars and volatiles are vaporizing. The reaction is self-sustaining (exothermic). **3. Blue/Transparent Smoke:** Carbonization complete. The wood is now charcoal. If you wait longer, the charcoal turns to ash.

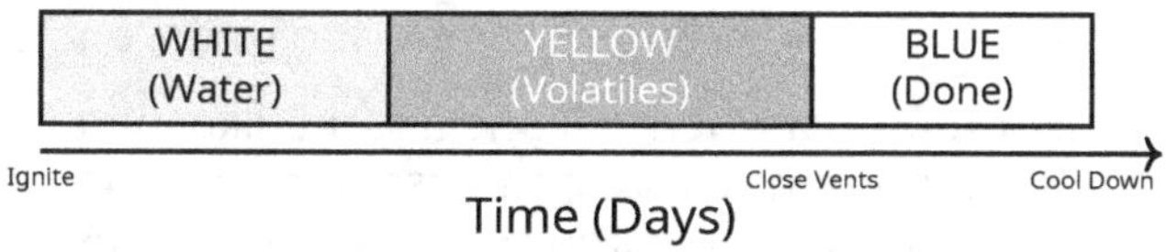

Figure 4: Phases of the burn cycle.

Harvesting

Once the smoke turns blue, seal the kiln completely with wet mud. No air must enter. The pile requires 24 to 48 hours to cool.

Warning: Opening the kiln while it is still hot (above 50°C) introduces oxygen to superheated fuel, causing the charcoal to spontaneously ignite and burn to ash instantly. Patience preserves the yield.

End of Discovery #009

DISCOVERY #010: CLAY & CERAMICS

Category: Materials / Chemistry **Status:** Essential

From Earth to Stone

Clay provides the ability to shape waterproof, heat-resistant containers from common dirt. This allows for boiling water, storing grain, and cooking food.

Clay acts as a plastic material when wet and a stone-like material when fired. This transformation occurs due to its microscopic structure. Clay consists of microscopic flat plates (alumina and silica). Water acts as a lubricant between these plates, allowing them to slide past each other (plasticity). Heat removes this water permanently, fusing the plates together.

Identifying Clay

Clay is found near riverbanks or exposed subsoil layers. **The Snake Test:** Roll a sample of damp earth into a pencil-thick rope. Wrap it around your finger.

- **Breaks immediately:** Silt or sand (Unusable).
- **Cracks slightly:** Poor clay (Usable with processing).
- **No cracks:** High-quality plastic clay.

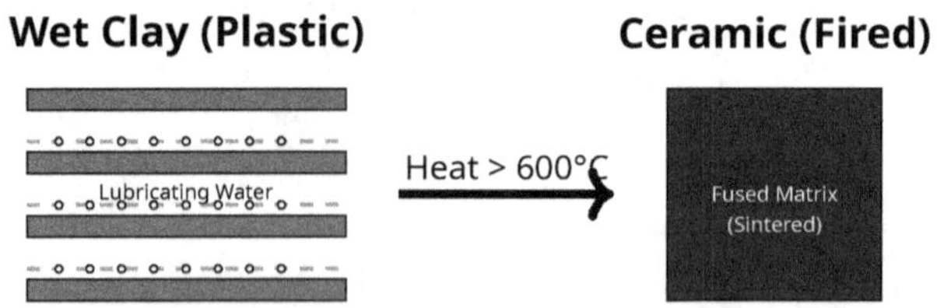

Figure 1: Molecular fusion removes chemical water.

Processing and Temper

Raw clay inevitably contains rocks and organic debris. **Levigation:** Mix clay with water to form a thin slurry (slip). Heavy rocks sink; organic matter floats. Pour off the middle layer containing pure clay into a drying pit.

The Necessity of Temper

Pure clay shrinks significantly (10-15%) when drying. This shrinkage causes warping and cracking. To prevent this, add **Temper**. Temper is non-plastic material that provides a rigid skeleton for the clay particles to cling to.

Common Temper Materials:

- **Sand:** Readily available.
- **Grog:** Fired pottery crushed into powder. Best for thermal shock resistance.
- **Crushed Shells:** Good binder, but decomposes at very high heat.

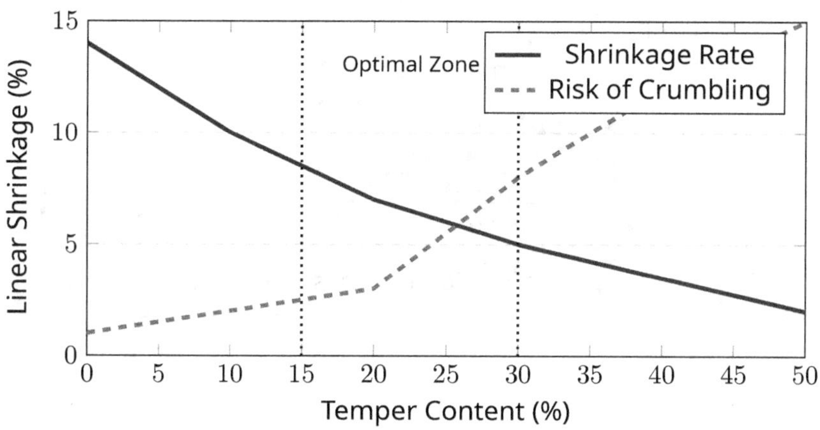

Figure 2: Balancing structural stability with plasticity.

The ideal ratio is typically 1 part temper to 3 or 4 parts clay (20-25% temper).

Vessel Construction

The Potter's Wheel accelerates production but is not required for survival. The **Coil Method** allows for the construction of vessels of any size, including large water storage jars.

The Coil Process:

1. **Base:** Form a flat disc of clay for the bottom.
2. **Rolling:** Roll clay into long, uniform ropes (snakes).
3. **Stacking:** Spiral the ropes upward to build the walls.
4. **Bonding:** Smear the clay downward on the inside and upward on the outside to fuse the coils. This mechanical bond is critical.

Drying

Drying must be slow and even. Rapid drying creates steam pressure inside the walls, causing explosions during firing. Dry pots in the shade, away from wind, for at least one week. The pot is ready when it feels room temperature to the touch (not cold).

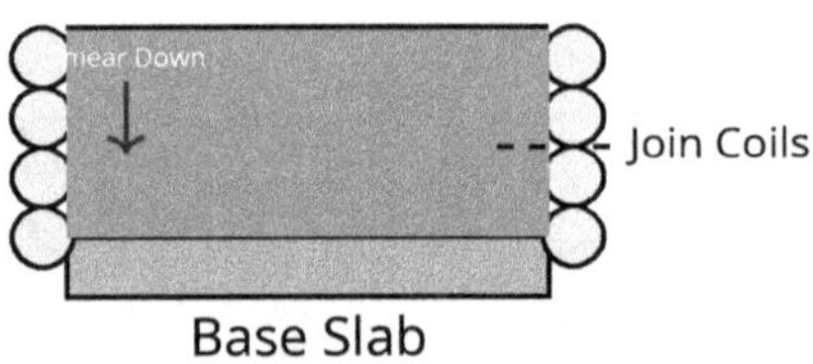

Figure 3: Cross-section: Joining coils to form a solid wall.

The Firing Process

Firing converts "Greenware" (dry clay) into "Bisqueware" (stone). This requires reaching at least 600°C, though 900°C is preferred for durability.

Phase 1: Water Smoking (20°C - 200°C) Remaining atmospheric water evaporates. Heat slowly to prevent steam explosions.

Phase 2: Quartz Inversion (573°C) Silica crystals expand suddenly. The temperature must rise evenly through this point to prevent cracking.

Phase 3: Sintering (600°C - 900°C) Chemical water is driven off. The clay particles fuse. The pot glows cherry red.

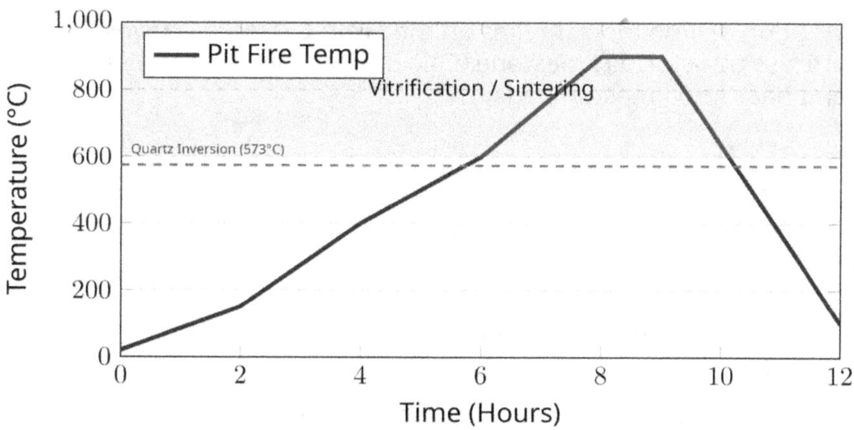

Figure 4: Temperature management avoids thermal shock.

Summary

Ceramics are the first synthetic material created by humanity. They provide rodent-proof grain storage, watertight vessels for transport, and heat-resistant pots for cooking, enabling the sterilization of water and the preparation of nutrient-dense stews.

End of Discovery #010

DISCOVERY #011: LIME & MORTAR

Category: Chemistry / Construction **Status:** Essential

The Chemistry of Stone

Dry stacked stones are unstable. To build permanent structures, you need a binding agent that conforms to irregular shapes and hardens into rock. Mud washes away in the rain. **Lime Mortar** endures for millennia.

The creation of lime mortar relies on the **Lime Cycle**. This reversible chemical process transforms limestone (Calcium Carbonate) into a reactive powder and back into limestone again.

The Three Phases

The cycle consists of three distinct chemical reactions:

1. **Calcination (Burning):** Heat drives Carbon Dioxide (CO_2) out of the stone.
2. **Hydration (Slaking):** Water is added to the burnt stone, creating a paste.
3. **Carbonation (Curing):** The paste absorbs CO_2 from the air, returning to its original chemical state as stone.

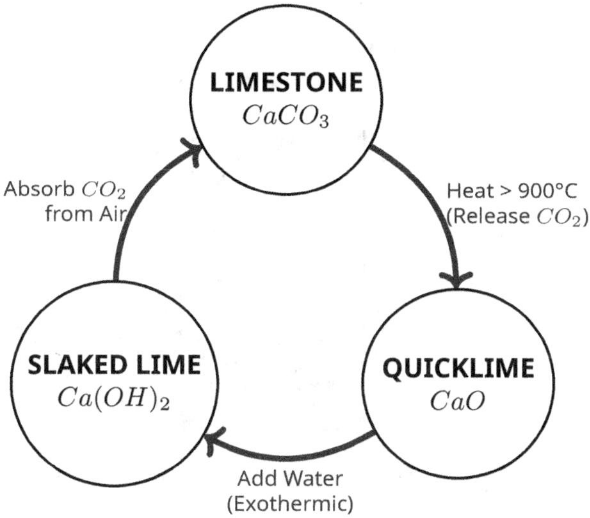

Figure 1: The eternal chemical loop of calcium.

Production: The Lime Kiln

The raw material is Limestone, Chalk, marble, or seashells (all Calcium Carbonate). To convert this stable rock into reactive Quicklime, you must heat it to at least 900°C.

The most efficient method is the **Flare Kiln** or **Draw Kiln**. The stone creates an arch over the fire, or is layered with fuel.

Visual Indicators: Limestone glows brightly when heated. As it converts to Quicklime, the stone becomes lighter (losing 44% of its weight as gas) and glows with a fierce white incandescence (the origin of the phrase "Limelight").

Safety Warning

Quicklime (CaO) is caustic. It burns skin and blinds eyes. When it contacts water, it reacts violently, boiling the water instantly. Always wear eye protection and handle dry quicklime with extreme care.

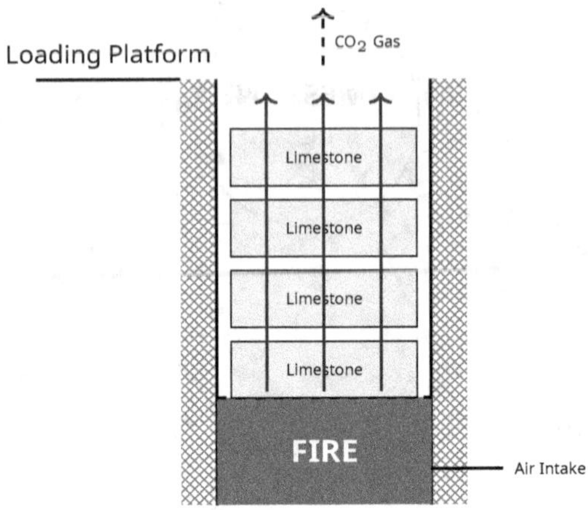

Figure 2: Continuous Draft Kiln construction.

Slaking and Mixing

Adding water to Quicklime produces **Slaked Lime** ($Ca(OH)_2$), a soft, white putty. This putty is the binder. However, pure lime paste shrinks and cracks when it dries.

To create functional **Mortar**, you must add **Aggregate** (Sand). The sand particles provide structural bulk and prevent shrinkage. The lime paste coats the sand grains, gluing them together.

The Golden Ratio: 1 part Lime Putty to 3 parts Sharp Sand. "Sharp" sand means angular grains (river sand), not round grains (beach/desert sand). Angular grains lock together physically.

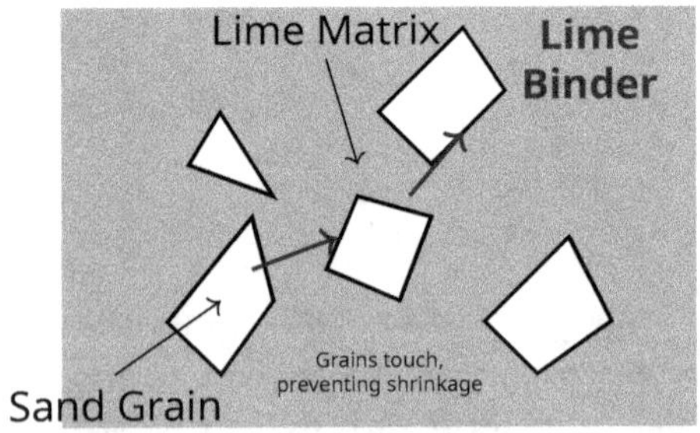

Figure 3: Microstructure: Aggregate provides the skeleton.

The Curing Process

Lime mortar does not "dry" like mud; it "cures" chemically. It absorbs Carbon Dioxide from the atmosphere to convert the Calcium Hydroxide back into Calcium Carbonate (Limestone).

The Time Factor: This reaction is slow. It proceeds from the surface inward at a rate of approximately 25mm (1 inch) per year.

- **Requirement:** The wall must be breathable. Sealing the surface with waterproof paint stops the carbonation process, causing the inner mortar to rot.
- **Strength:** The mortar continues to harden for centuries.

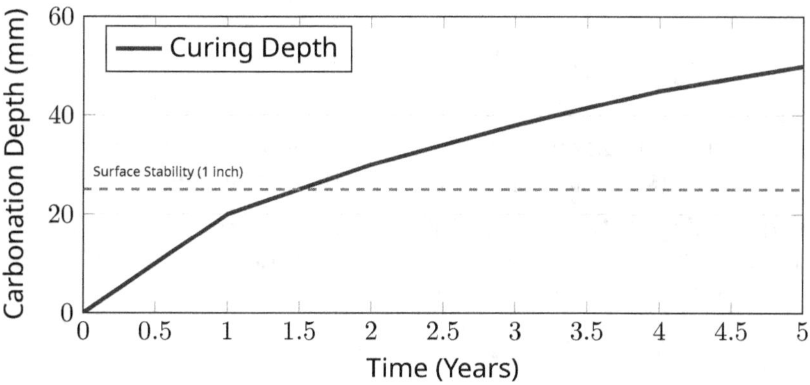

Figure 4: Slow reaction rate requires porous surfaces.

Summary

Lime mortar allows for the construction of vertical walls that withstand wind and rain. It is a chemical adhesive that turns separate stones into a monolithic structure, returning the dust of the earth back into the rock from which it came.

End of Discovery #011

DISCOVERY #012: FIRED BRICKS

Category: Construction / Materials **Status:** Essential

Artificial Stone

Fired clay bricks provide the compressive strength of stone combined with the versatility of a manufactured product. The firing process creates a ceramic bond impervious to rain, frost, and fire. This durability enables the construction of multi-story infrastructure in regions lacking natural rock deposits.

Dimensional Standards

A brick functions as a standardized modular unit. Its dimensions allow for single-handed manipulation, optimizing labor efficiency.

The Interlocking Ratio: Successful masonry requires bricks to overlap. To turn a corner or terminate a wall without cutting, the length of the brick must equal twice its width plus the thickness of the mortar joint.

$$Length = 2 \times Width + Joint$$

A common standard follows the 4:2:1 ratio (Length:Width:Height), such as 215mm x 102.5mm x 65mm.

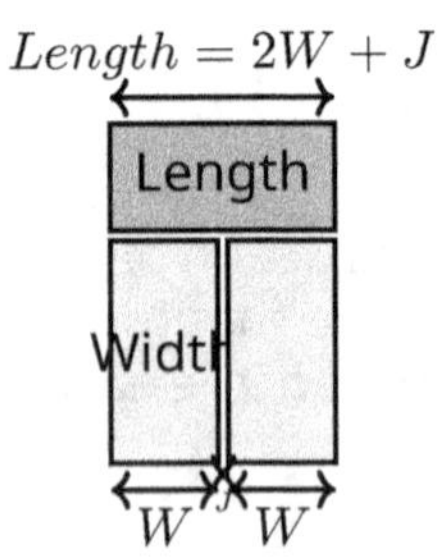

Figure 1: Standardization enables interlocking corners.

Molding and Processing

Brick production begins with tempering the clay. Sand is added to reduce plasticity and shrinkage. The mixture is thrown into a wooden frame (mold) dusted with sand to prevent sticking.

Drying Phase: Raw "green" bricks contain physical water. Firing wet bricks causes steam explosions. Bricks are stacked in a "hack" formation under shelter for 2 to 4 weeks. Air circulates around each brick, removing moisture until the brick is leather-hard.

Shrinkage Calculation: Clay shrinks during drying and again during firing (sintering). Total volume loss can reach 15%. Molds are constructed 10-15% larger than the required final dimensions to compensate for this reduction.

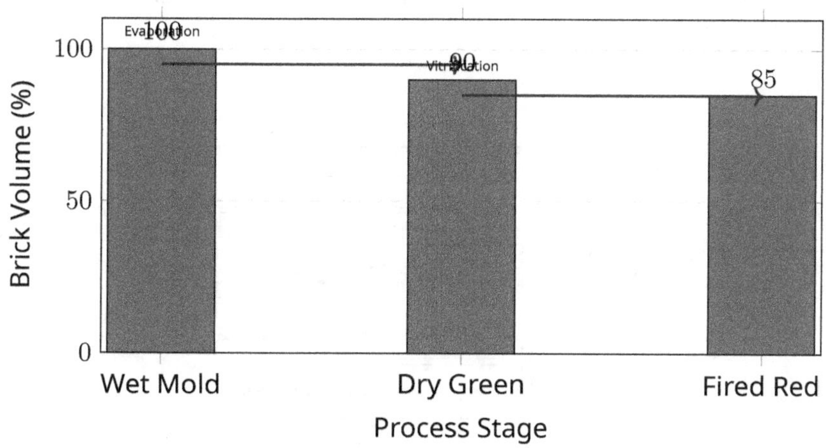

Figure 2: Volume loss requires oversized molds.

The Clamp Kiln

A Clamp Kiln uses the bricks themselves to form the furnace structure. This method allows for mass production without prior infrastructure.

Structure:

- **Base:** Fired bricks protect the stack from ground moisture.
- **Fuel Layers:** Coal dust or charcoal is packed between layers of green bricks.
- **Insulation:** The entire stack is encased in mud (pug) to retain heat.

The fire burns slowly from the bottom up, taking weeks to complete. The bricks in the center are fired hardest (clinkers), while outer bricks may remain underfired (commons).

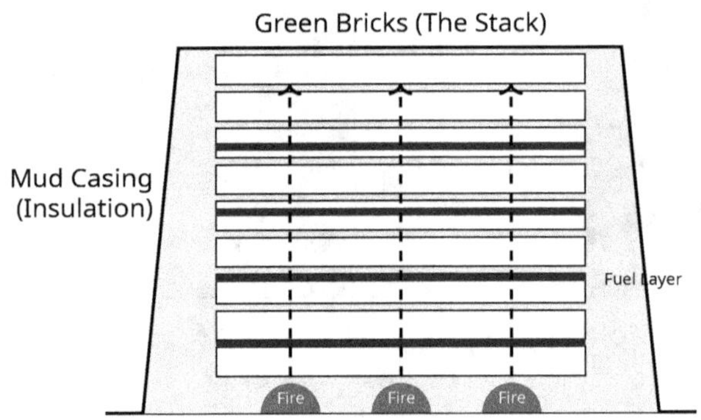

Figure 3: The Clamp: The product forms the furnace.

Distribution of Force

A brick wall functions as a single monolithic unit due to the bonding pattern. The mortar levels the bed and seals the joint. Gravity and friction provide the primary structural stability.

The Rule of Bonding: Vertical joints must never align between courses. Aligned joints create a weak point where cracks can propagate vertically. By overlapping bricks, a point load applied to the top is distributed in a triangular pattern through the wall, reducing the stress on any individual brick.

English Bond: This strong bond alternates rows of headers (ends) and stretchers (sides), tying the face of the wall to the core.

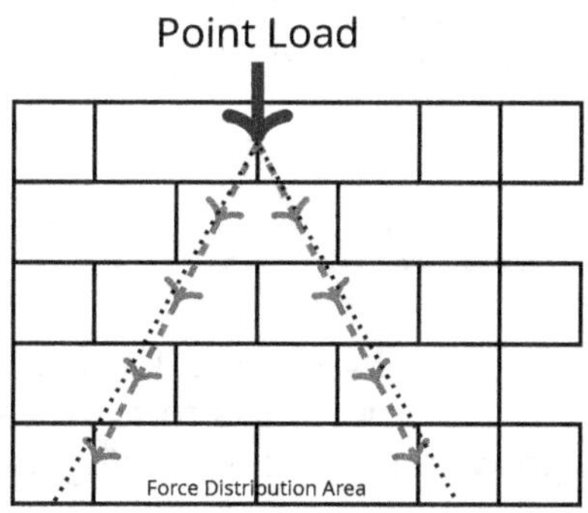

Figure 4: Overlapping joints distribute point loads.

Summary

Fired bricks convert a variable resource (earth) into a standardized, imperishable component. This modularity allows for the rapid construction of complex structures, from aqueducts to factories, using simple arithmetic and skilled labor.

End of Discovery #012

DISCOVERY #013: CEMENT

Category: Construction / Chemistry **Status:** Essential

Liquid Stone

Standard lime mortar requires exposure to air (Carbon Dioxide) to cure. This limitation prevents it from hardening underwater or in the center of thick walls. Infrastructure such as dams, piers, and heavy fortifications requires a **Hydraulic Binder**.

Hydraulic cement sets through a reaction with water, not air. This chemical process transforms a paste into a solid, water-insoluble rock, regardless of the environment.

The Chemistry of Hardening

The hardening process involves hydration. The calcium silicates in the cement powder react with water molecules to form calcium silicate hydrate crystals. These crystals grow and interlock, locking the aggregate (sand/gravel) into a rigid matrix.

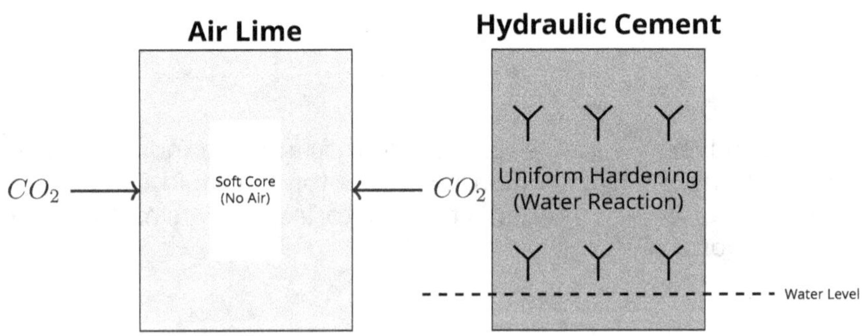

Figure 1: Hydraulic binders harden uniformly, even underwater.

Roman Concrete (Opus Caementicium)

The Romans built the Pantheon and harbor moles using a natural hydraulic cement. They discovered that mixing lime putty with volcanic ash created a superior binder.

Pozzolana: Volcanic ash contains reactive silica and alumina. When mixed with lime and water, the silica reacts with the calcium hydroxide to form calcium silicate hydrate (C-S-H). This is the **Pozzolanic Reaction**.

Artificial Pozzolana: In the absence of volcanic ash, crushed brick or tile dust serves as a substitute. The firing of the clay in the brick activates the silica, allowing it to react with the lime.

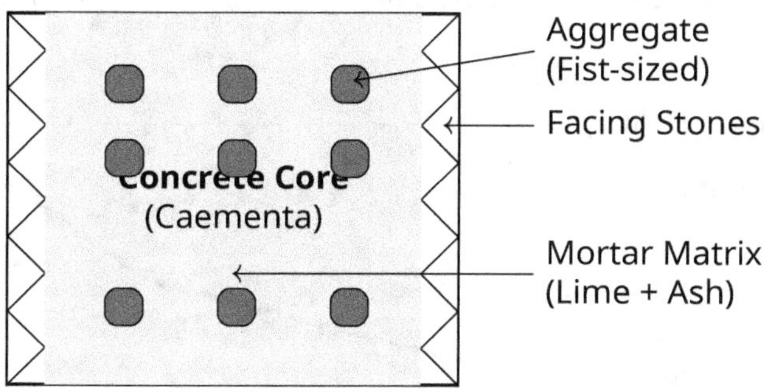

Figure 2: Roman walls use cement to bind random rubble.

Artificial Cement (Portland)

Without volcanic ash, we manufacture hydraulic cement by heating limestone and clay together. This creates **Portland Cement**, the standard for modern infrastructure.

The Recipe:

1. **Limestone ($CaCO_3$):** Provides Calcium.
2. **Clay/Shale ($SiO_2 + Al_2O_3$):** Provides Silica and Alumina.

The materials are ground to powder and heated to 1450°C. At this temperature, the materials partially melt and fuse chemically, forming small nodules called **Clinker**. The clinker is cooled and ground into a fine powder.

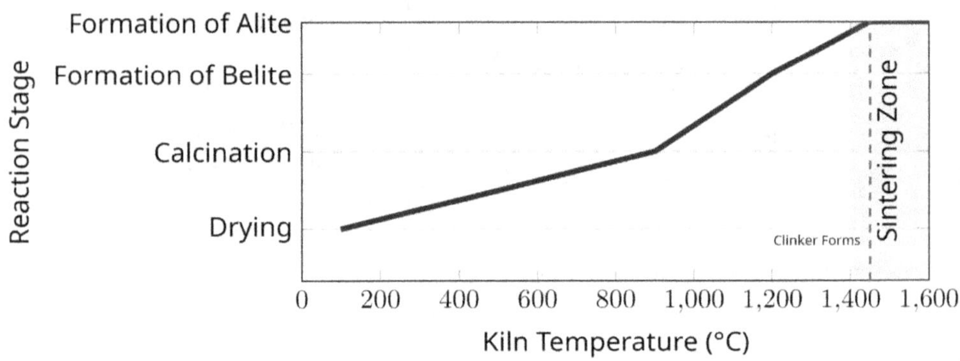

Figure 3: High heat creates new chemical compounds (Alite).

Gypsum

A small amount (5%) of Gypsum is added during the final grinding. This retards the setting time, preventing the cement from hardening instantly upon contact with water ("flash set").

The Composite Era

Cement is the glue; **Concrete** is the rock. Concrete consists of Cement (10-15%), Water (15-20%), and Aggregates (65-75%).

Concrete possesses immense compressive strength (squeezing) but low tensile strength (pulling). It cracks under tension. Steel possesses high tensile strength. Combining the two creates **Reinforced Concrete**.

The coefficient of thermal expansion for steel and concrete is identical. They expand and contract together, ensuring the bond remains intact through seasons.

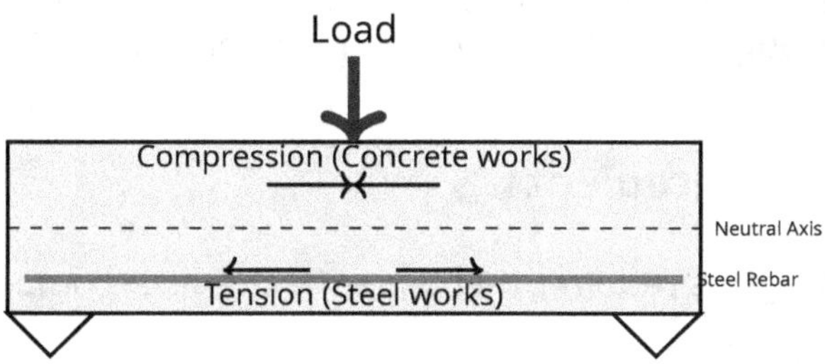

Figure 4: Steel takes the tension; Concrete takes the compression.

Summary

Cement transforms geology from a constraint into a choice. By mastering the chemistry of calcium silicates, you build structures that defy water and time, laying the literal foundation for a permanent civilization.

End of Discovery #013

DISCOVERY #014: GLASS

Category: Materials / Chemistry **Status:** Essential

Invisible Matter

Glass is a unique state of matter. It possesses the mechanical rigidity of a solid and the disordered molecular structure of a liquid. This "amorphous solid" state grants glass its transparency and chemical inertness. It allows light to pass through and blocks air, water, and bacteria.

The Trinity of Ingredients

Producing viable glass requires three specific chemical components melted together.

1. The Former (Silica): Sand (SiO_2). This is the body of the glass. Pure silica melts at a prohibitive 1713°C. **2. The Flux (Soda Ash):** Sodium Carbonate (Na_2CO_3). This lowers the melting point of silica to a manageable 1000°C. This addition renders the glass water-soluble (water glass). **3. The Stabilizer (Lime):** Calcium Oxide (CaO). Adding lime makes the glass insoluble in water and chemically durable.

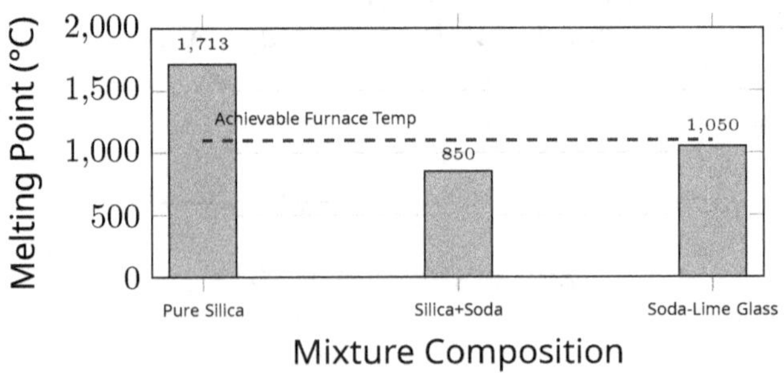

Figure 1: The Flux makes melting sand possible.

The Glass Furnace

The furnace must sustain temperatures above 1000°C for extended periods.

The Pot Furnace: Clay crucibles (pots) sit on a shelf inside the furnace. The fire burns below, and hot gases circulate around the pots. This separates the glass from the ash and fuel impurities, ensuring clarity.

Fining: Molten glass contains gas bubbles. Maintaining high heat allows these bubbles to rise to the surface. Adding fining agents (like Arsenic or Antimony, if available) accelerates this clarification.

Annealing: Hot glass shrinks as it cools. If the surface cools faster than the core, stress builds up, causing the glass to shatter. **Annealing** involves cooling the glass very slowly in a separate oven (the Lehr) to equalize internal stress.

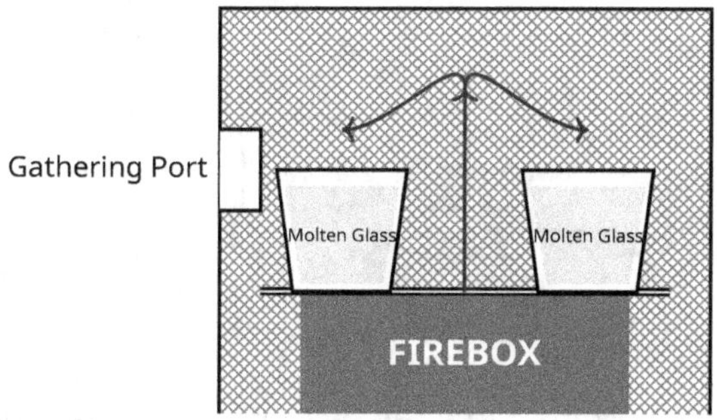

Figure 2: Indirect heat protects glass from ash contamination.

Forming Techniques

Glass is shaped while in its plastic state (800°C - 1000°C).

Crown Glass (Windows): To make a flat pane from a round bubble: 1. Blow a sphere on the blowpipe. 2. Transfer the sphere to a solid iron rod (punty) attached opposite the blowpipe. 3. Crack off the blowpipe, leaving an opening. 4. Spin the punty rapidly. Centrifugal force flattens the open sphere into a large, flat disc. 5. Cut the disc into panes. The center (bullseye) is thickest.

Mold Blowing (Bottles): Blow the bubble into a wooden or metal mold. This standardizes volume and shape, essential for scientific measurements and trade.

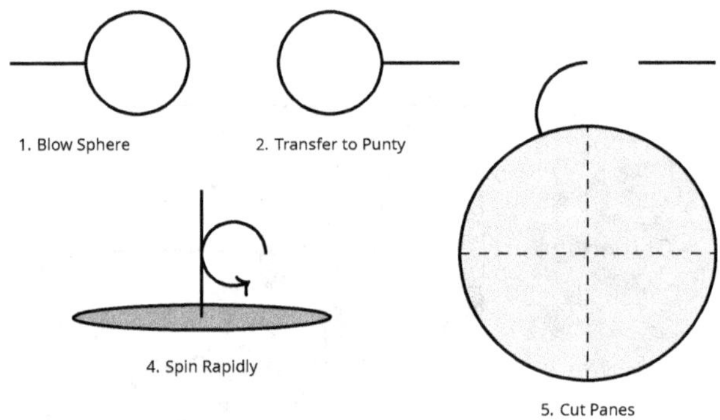

Figure 3: Centrifugal force creates flat glass sheets.

The Transparent Revolution

Glass has two properties that change civilization: Chemical Inertness and Refraction.

Chemical Inertness: Glass does not react with acids or foods. It allows for the sterile storage of medicines, the pickling of food, and sophisticated chemical synthesis. It is non-porous and easy to clean.

Optics: Glass slows down light passing through it, causing the light beam to bend (**Refraction**). By grinding a piece of clear glass into a convex curve, you focus parallel light rays to a single point. This is the basis of the lens.

- **Spectacles:** Extend the working life of craftsmen.
- **Microscopes:** Reveal the germ theory of disease.
- **Telescopes:** Reveal the mechanics of the universe.

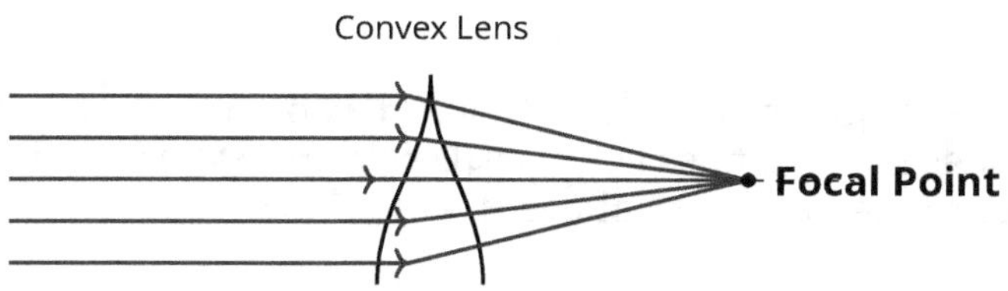

Figure 4: Controlling light enables science and medicine.

Summary

Glass is more than a building material; it is the material of observation. It provides the window through which we observe the microscopic and the cosmic, and the vessel in which we preserve the chemistry of life.

End of Discovery #014

DISCOVERY #015: COPPER & BRONZE

Category: Metallurgy / Materials **Status:** Essential

The First Metal

Stone tools are brittle. Metal tools are ductile and reparable. Copper is the gateway to metallurgy because it is occasionally found in its native metallic state and has a relatively low melting point compared to iron.

Copper is extracted from green rocks like Malachite ($CuCO_3Cu(OH)_2$). Heating these rocks in a charcoal fire triggers a chemical reduction, releasing pure copper metal.

Thermal Requirements

Melting copper requires 1085°C. A standard wood fire reaches only 700°C. To smelt copper, you must use charcoal (Discovery #009) and forced air (bellows) to raise the temperature above the melting point.

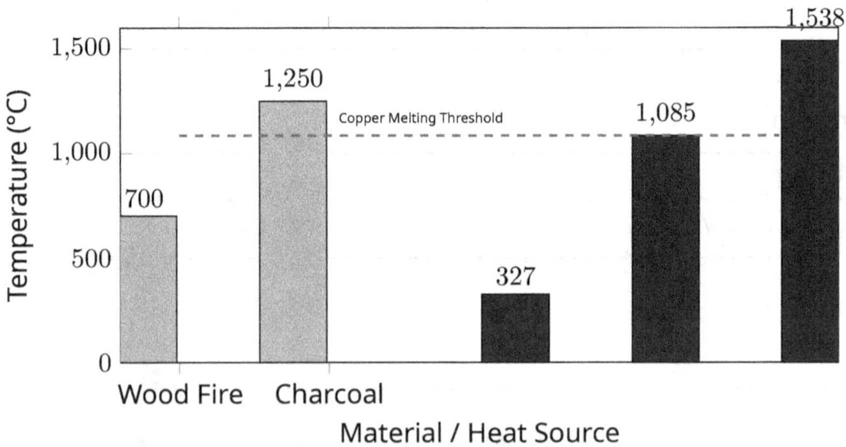

Figure 1: Why charcoal is mandatory for the Copper Age.

The Smelting Process

Smelting converts rock into metal. It is a chemical reaction, not just melting. **Reduction:** The carbon in the charcoal steals oxygen atoms from the copper ore, leaving pure copper behind.

$$2CuO + C \rightarrow 2Cu + CO_2$$

The Bowl Furnace: 1. Dig a small hole in the ground. 2. Line it with clay. 3. Fill with charcoal and crushed green ore (Malachite). 4. Pump air into the base using a tuyere (clay pipe) and bellows. 5. Maintain heat for hours. The copper will melt and sink to the bottom, forming a "button" or ingot.

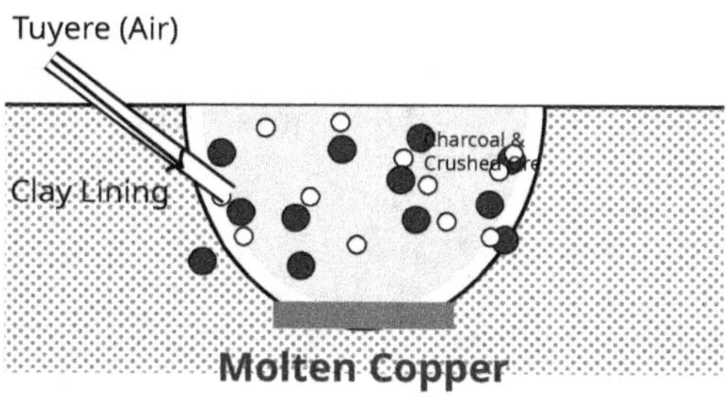

Figure 2: Simple reduction furnace using forced air.

The Birth of Bronze

Pure copper is soft. It dulls quickly when used as a knife or axe. To create a durable tool, you must create an **Alloy**.

Bronze is a mixture of Copper (90%) and Tin (10%). Tin atoms are larger than copper atoms. When introduced into the copper crystal lattice, they strain the structure. This atomic distortion prevents the metal layers from sliding over each other, making the alloy significantly harder than either parent metal.

Sources: Tin is rare (found as Cassiterite, heavy black stones). Arsenic can also be used (Arsenical Bronze) but is toxic to the smith.

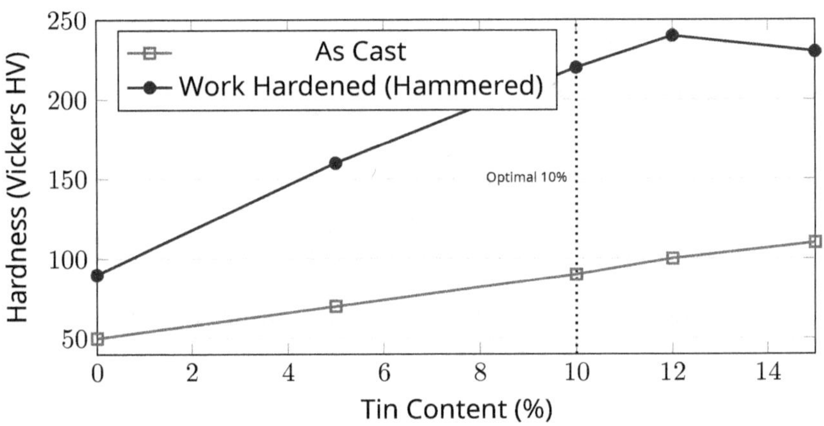

Figure 3: Alloying plus hammering triples the hardness.

Work Hardening

Bronze edges are sharpened by hammering (cold forging). This compresses the crystal structure further, creating a razor edge capable of cutting wood or flesh.

Shaping Metal

Metal possesses the unique property of fusibility. It can be melted and poured into complex shapes. This is **Casting**.

Types of Molds:

- **Open Mold (Stone):** A shape carved into a flat rock. Produces items flat on one side (ingots, simple axes).
- **Bivalve Mold (Stone/Clay):** Two halves pinned together. Allows for 3D shapes and sockets for handles.
- **Sand Casting:** Pressing a pattern into damp, oil-bound sand. Disposable but infinite versatility.

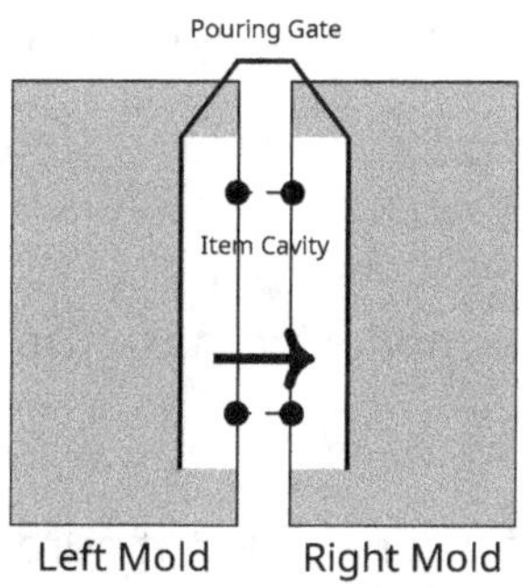

Figure 4: Two-part molds allow for complex 3D tools.

Summary

Copper allows for the creation of water-tight vessels and basic conductors. Bronze provides the first durable tools for carpentry, warfare, and agriculture, surpassing the limitations of stone and bone.

End of Discovery #015

DISCOVERY #016: THE BLOOMERY

Category: Metallurgy / Materials **Status:** Essential

The Iron Paradox

Iron is the most abundant metal on Earth, yet it was the last to be mastered. The difficulty lies in the temperature. Iron melts at 1538°C. A charcoal furnace typically reaches 1200-1300°C. You cannot melt iron with simple technology.

The solution is the **Bloomery Process**. This method extracts iron without melting it. It relies on chemical reduction in the solid state. The iron remains solid, while the waste rock (slag) melts and drains away.

Solid State Reduction

Inside the furnace, carbon monoxide gas strips oxygen atoms from the solid iron ore.

$$Fe_2O_3 + 3CO \rightarrow 2Fe + 3CO_2$$

This leaves behind microscopic particles of pure iron. These particles weld themselves together into a spongy mass called a "Bloom."

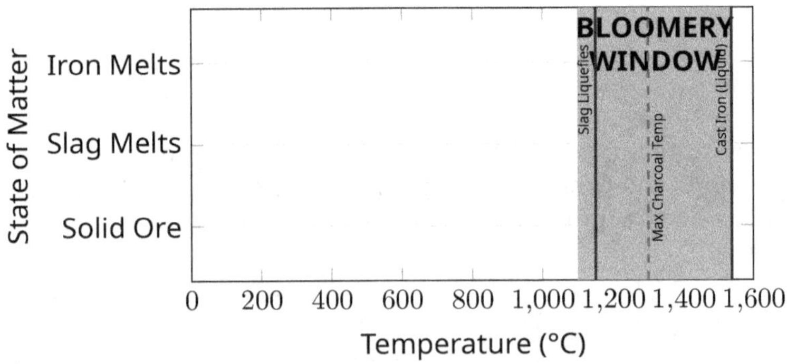

Figure 1: Iron is produced in the narrow gap between 1150°C and 1538°C.

Furnace Architecture

A bloomery is a vertical shaft furnace constructed from clay and straw (adobe). It acts as a counter-current reactor: fuel and ore go down, hot gases go up.

Dimensions:

- **Height:** 1.0 to 1.5 meters. Height creates the draft.
- **Internal Diameter:** 30-40 cm. Too wide, and the air cannot penetrate to the center.
- **Walls:** Thick (15-20 cm) to insulate the reaction.

Air Supply: A clay pipe (**Tuyere**) enters near the base. Bellows pump air through this pipe into the reaction zone.

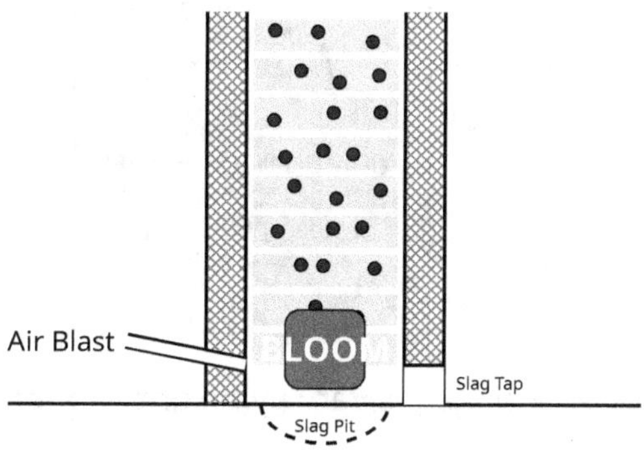

Figure 2: The shaft furnace concentrates heat and gas.

The Operational Cycle

A smelt requires continuous labor for 8 to 12 hours.

1. Pre-heating: Start a wood fire to dry the furnace walls. Fill with charcoal only. **2. Charging:** Once the furnace is hot (yellow heat), add layers of charcoal and roasted ore in a 1:1 ratio by weight. **3. The Blast:** Operate the bellows rhythmically. The goal is a steady 1200°C. **4. Formation:** As the charge descends, the ore reduces to iron. The silicate rock melts into liquid slag. The heavy liquid slag trickles down, bypassing the solid iron sponge forming near the air blast.

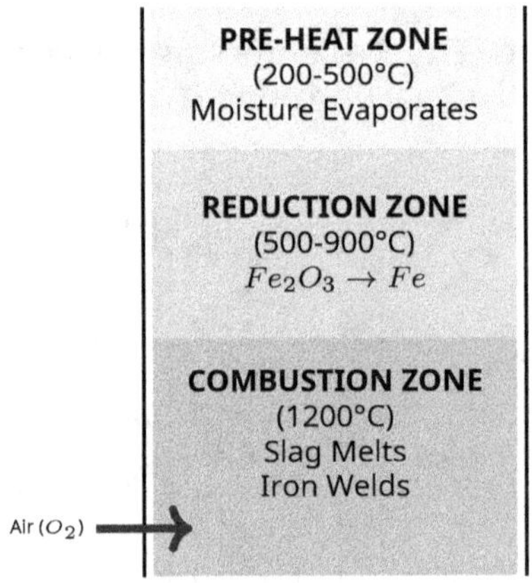

Figure 3: Chemical stages vary by height and temperature.

Slag Management

If the slag does not drain, it will drown the iron sponge and freeze the air inlet. You must periodically poke a hole in the bottom arch to let the "lava" flow out.

From Sponge to Bar

At the end of the burn, break down the furnace wall. You will find a glowing, dirty lump the size of a human head. This is the **Bloom**.

It is not yet usable metal. It is a matrix of iron and glass (slag). **Consolidation:** While the bloom is white-hot, drag it to an anvil (a large flat stone). Hammer it with heavy wooden mallets. This mechanical compression squeezes the liquid slag out of the pores, like water from a sponge. The pure iron particles weld together into a solid mass of **Wrought Iron**.

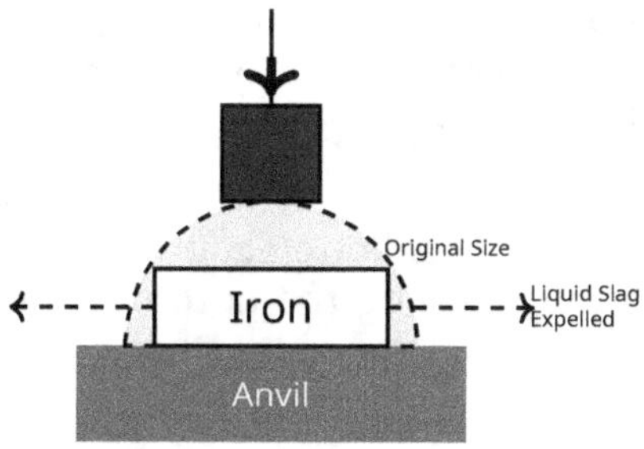

Figure 4: Refining the bloom: Mechanical purification.

Summary

The Bloomery allows the production of iron with temperatures attainable by simple technology. It yields wrought iron: a tough, fibrous metal that resists shock and tension, ideal for nails, hinges, tires, and tools.

End of Discovery #016

DISCOVERY #017: THE BLAST FURNACE

Category: Metallurgy / Industry **Status:** Essential

Liquid Iron

The Bloomery produces iron in kilogram batches. To build railways, bridges, and engines, civilization requires iron in tonnage. The Blast Furnace answers this need by operating continuously for months or years, producing a constant stream of liquid metal.

The Blast Furnace differs fundamentally from the Bloomery. It is designed to melt the iron completely. This liquefaction allows for the separation of metal and slag by density, enabling the automatic purification of massive quantities of ore.

The Vertical Reactor

The furnace creates a tall column of reacting materials. The height serves two functions. First, it pre-heats the descending charge, maximizing fuel efficiency. Second, it provides a long exposure time to reducing gases (CO), ensuring complete reduction of the ore before it reaches the melting zone.

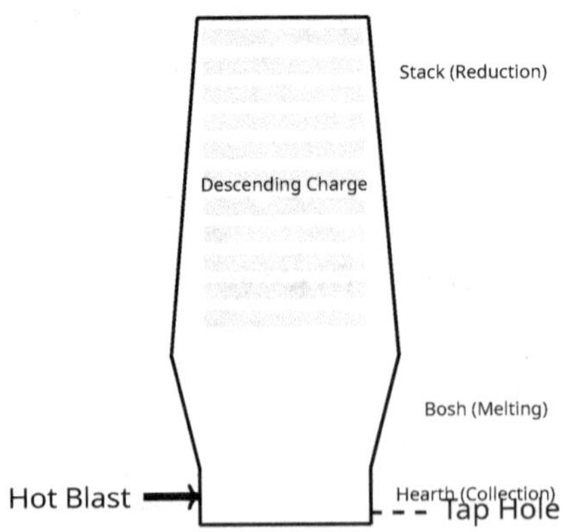

Figure 1: Continuous counter-current reactor design.

The Eutectic Advantage

Pure iron melts at 1538°C, a temperature difficult to maintain with charcoal. The Blast Furnace circumvents this by producing an alloy: **Cast Iron**.

In the high-temperature zone, the iron absorbs large amounts of carbon (3% to 4.5%) from the fuel. This creates an iron-carbon alloy with a significantly lower melting point of 1150°C. This "Eutectic Point" allows the metal to liquefy easily within the temperature range of a charcoal fire.

Carbon Absorption: The long descent through the charcoal stack saturates the iron with carbon. This turns the metal into a liquid that flows like water.

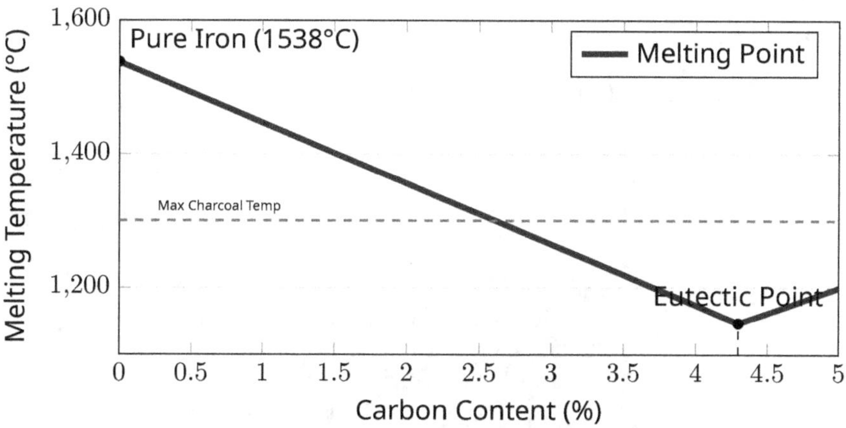

Figure 2: Carbon acts as a flux, lowering the melting point.

Slag Chemistry

Iron ore contains silica (sand) impurities. Silica creates a viscous, sticky slag that clogs the furnace. To fix this, we add a **Flux**: Limestone ($CaCO_3$).

The limestone decomposes into lime (CaO). The lime reacts with the silica (SiO_2) to form Calcium Silicate. This compound has a low melting point and low viscosity. It melts into a fluid liquid that floats on top of the molten iron, protecting it from re-oxidation.

The Charge Ratio: A typical charge consists of distinct layers dumped into the top:

1. **Fuel:** Charcoal or Coke (4 parts).
2. **Ore:** Iron Oxide (2 parts).
3. **Flux:** Limestone (1 part).

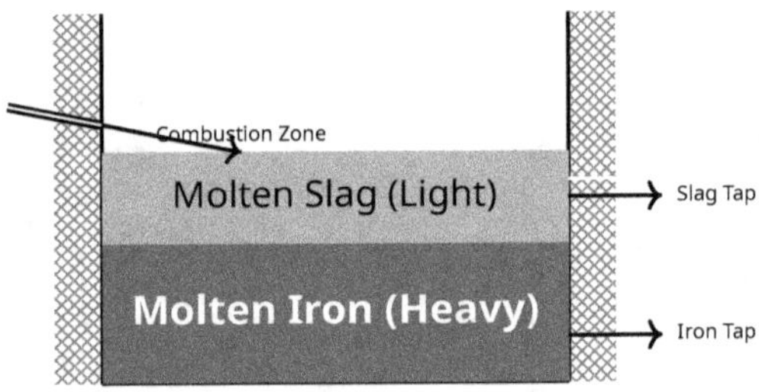

Figure 3: Density separation allows for purification.

Pig Iron Production

When the furnace is tapped, a river of white-hot metal flows out. It is directed into sand molds on the casting floor.

The layout resembles a nursing sow suckling piglets. The main channel is the "Sow," and the smaller side branches are the "Pigs." This gives the raw material its name: **Pig Iron**.

Properties: Pig iron is hard, wear-resistant, but brittle. It cannot be forged (hammered) or it will shatter. It is used for casting engine blocks, pipes, and cookware, or it is remelted and refined into steel (Discovery #018).

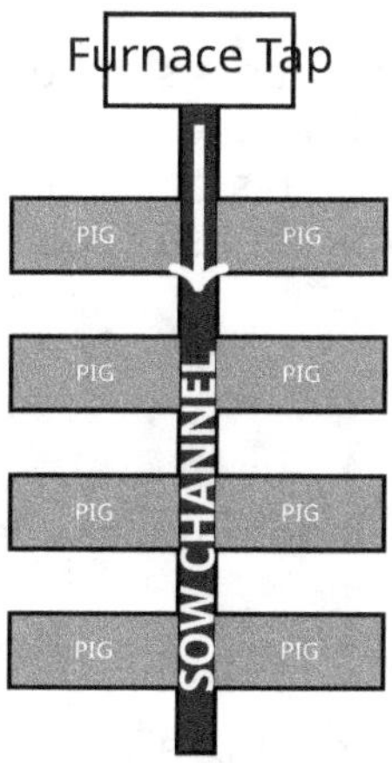

Figure 4: The Casting Floor: Efficient liquid management.

Summary

The Blast Furnace provides the volume of metal necessary for an industrial economy. By mastering liquid iron, you move from crafting individual tools to manufacturing machines, rails, and structural beams.

End of Discovery #017

DISCOVERY #018: STEEL

Category: Metallurgy / Industry **Status:** Essential

The Perfect Alloy

Iron exists in two basic forms: Wrought Iron (pure, soft) and Cast Iron (carbon-rich, brittle). Steel is an alloy of iron containing between 0.2% and 2.1% carbon.

This specific chemical balance grants steel unique mechanical properties. It possesses the hardness to hold a cutting edge and the elasticity to absorb shock without breaking. This combination defines the material requirements for springs, swords, and structural beams.

The Carbon Spectrum

Controlling the carbon content determines the metal's utility.

- **Low Carbon (Mild Steel):** Structural shapes, nails.
- **Medium Carbon:** Axles, gears, rails.
- **High Carbon:** Springs, knives, tools.

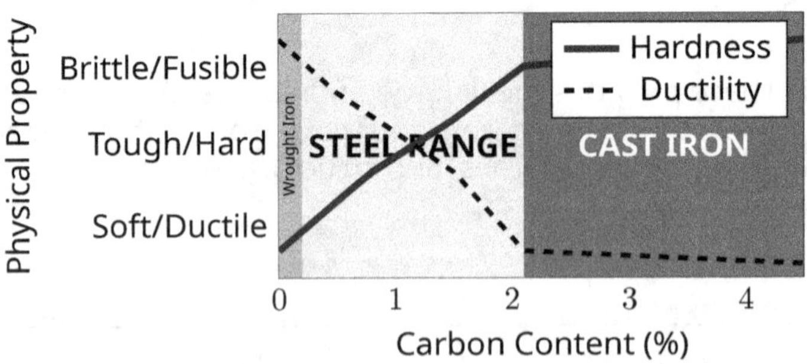

Figure 1: Steel balances hardness and ductility.

Method 1: Adding Carbon

Before industrialization, smiths created steel by adding carbon to wrought iron. This is the **Cementation Process**.

Iron bars are packed into stone chests filled with charcoal dust. The chest is sealed with clay to exclude oxygen and heated to red heat (1100°C) for 7 to 10 days. The carbon atoms diffuse into the solid iron lattice, transforming the outer layers into steel.

Blister Steel: The surface of the resulting bars is covered in blisters from escaping gas. To make the steel uniform, these bars are bundled, heated, and hammered together (Forge Welding) to distribute the carbon evenly.

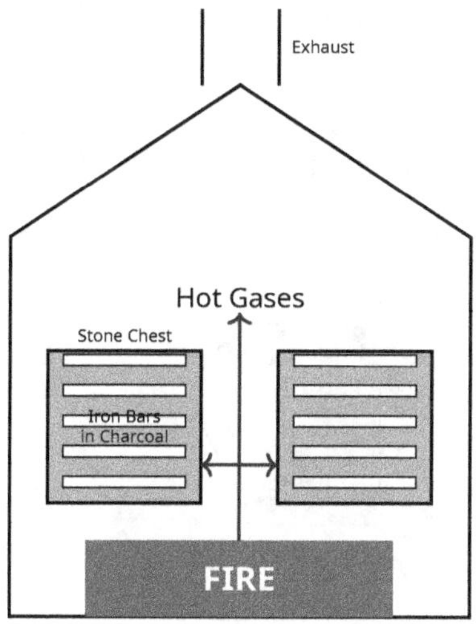

Figure 2: Solid-state diffusion of carbon (slow method).

Method 2: Removing Carbon

Industrialization requires tonnage. The **Bessemer Process** creates steel by removing excess carbon from liquid pig iron.

Pig iron contains 4% carbon. By blowing atmospheric air through the molten metal, the oxygen reacts with the carbon to form Carbon Monoxide (CO). This gas burns violently at the mouth of the converter. This reaction is exothermic, raising the temperature of the iron without additional fuel.

The Converter: A tilting vessel lined with refractory brick. Air enters through tuyeres at the bottom.

1. **Blow:** Air oxidizes silicon, manganese, and carbon.
2. **Flame:** A massive flame shoots from the mouth. When the flame drops, the iron is pure.
3. **Recarburization:** A precise amount of carbon (Spiegeleisen) is added back to reach the desired steel grade.

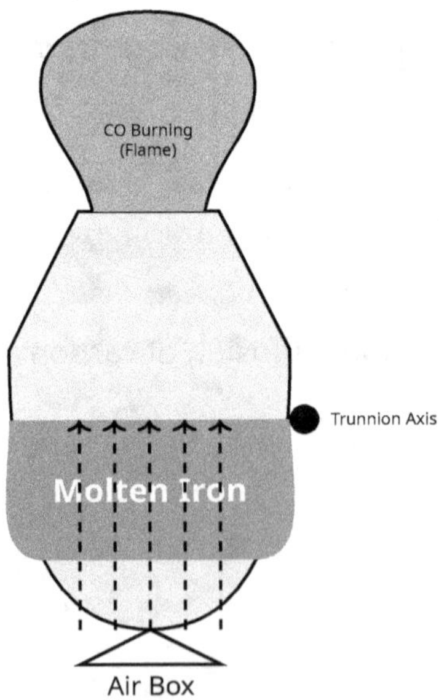

Figure 3: Oxidation converts Pig Iron into Steel in 20 minutes.

Tempering Steel

Steel's properties change depending on how it cools. **Quenching:** Plunging red-hot steel (Austenite) into water freezes the carbon atoms in a stressed crystal structure called **Martensite**. This creates maximum hardness but maximum brittleness. A quenched file will snap if dropped.

Tempering: Re-heating the quenched steel to a lower temperature (200-300°C) relaxes the crystal stress. This sacrifices some hardness to regain toughness.

Oxide Colors: The color of the oxide film on the steel surface indicates the temperature.

- **Straw (230°C):** Hardest. Razors, drills.
- **Purple (270°C):** Balanced. Knives, swords.
- **Blue (300°C):** Toughest. Springs, saws.

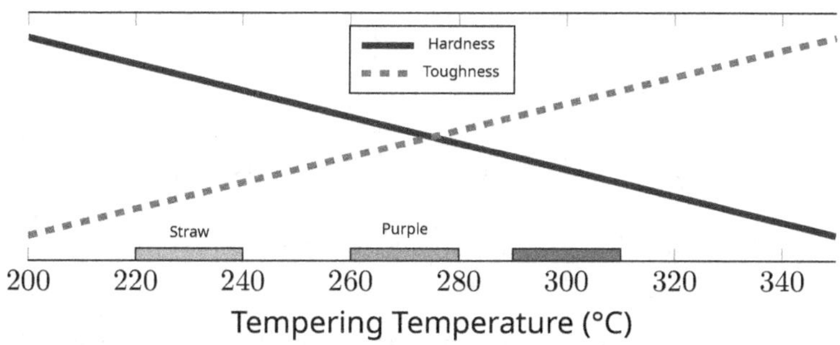

Figure 4: Select the temperature to match the tool's purpose.

Summary

Steel is the master material of industry. Cementation provides high-quality steel for tools. The Bessemer process provides bulk steel for rails and beams. Heat treatment tailors the metal to the specific mechanical demands of the application.

End of Discovery #018

DISCOVERY #019: WIRE & ROLLING

Category: Manufacturing / Metallurgy **Status:** Essential

The Geometry of Metal

Casting produces complex shapes with porous, brittle crystal structures. To create uniform, high-strength components, the metal requires mechanical working.

Plastic deformation rearranges the internal crystal lattice of the metal without removing material. By applying force exceeding the metal's yield strength, the material changes shape permanently. This process, known as wrought working, aligns the grain structure, increasing toughness and directional strength.

Stress and Strain

Successful metalworking operates within the "Plastic Region."

- **Elastic Region:** The metal acts like a spring. It returns to its original shape when force is removed.
- **Plastic Region:** The metal flows like clay. It holds the new shape.
- **Fracture Point:** The metal breaks.

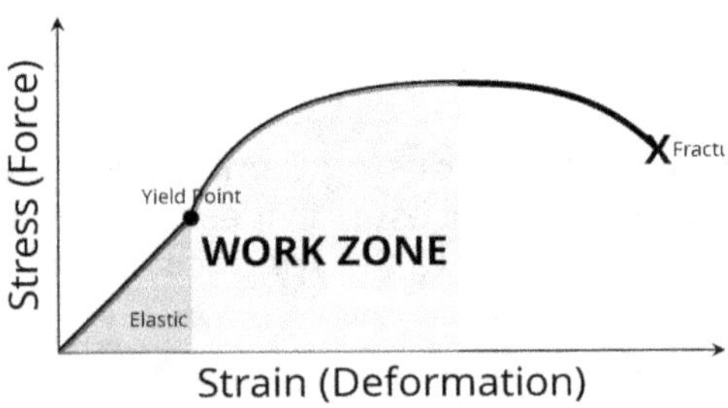

Figure 1: Shaping occurs between the Yield Point and Fracture.

The Rolling Mill

Rolling reduces the thickness of a metal ingot by passing it between two heavy, rotating cylinders. Friction pulls the metal into the gap, and compressive force thins it.

Because the volume of the metal remains constant, a reduction in thickness results in a proportional increase in length. The grain of the metal elongates in the direction of rolling, creating "grain flow" that resists cracking.

Hot vs. Cold Rolling:

- **Hot Rolling:** Performed above recrystallization temperature. Easy deformation for large structural beams.
- **Cold Rolling:** Performed at room temperature. Increases hardness and produces a smooth, precise finish for sheets and coins.

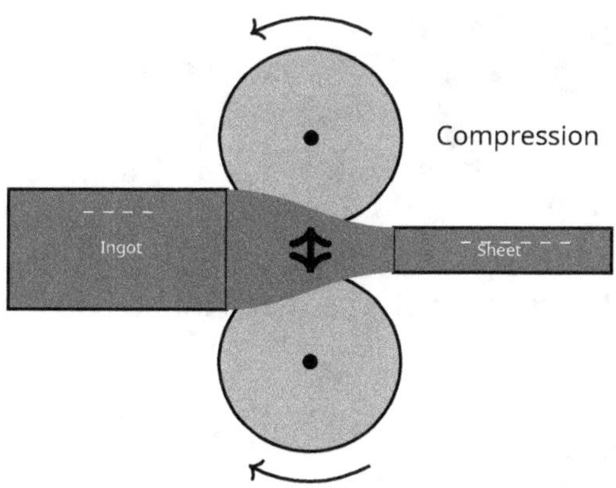

Figure 2: Volume is conserved: thickness decreases, length increases.

The Draw Plate

Wire is essential for nails, rivets, chainmail, and telegraphy. It is produced by pulling (drawing) a metal rod through a tapered hole in a hard steel plate.

The hole is smaller than the rod. As the metal passes through, the radial compressive force of the die wall reduces its diameter.

Die Geometry:

- **Entry Angle:** Guides the wire and holds lubricant (tallow/wax).
- **Bearing:** The narrowest cylindrical section that sizes the wire.
- **Relief Angle:** Allows the wire to exit without scraping.

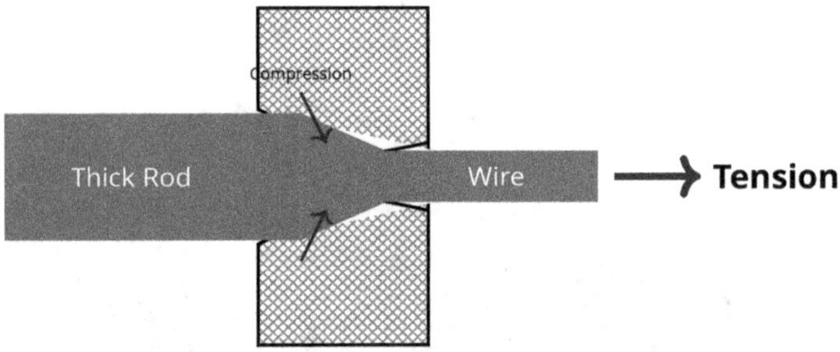

Figure 3: The die funnels the metal into a narrower gauge.

Annealing Steps

Drawing hardens the metal (work hardening). If drawn too far, it snaps. The wire must be periodically heated (annealed) to relax the crystal structure before drawing it through the next smaller die.

The Draw Bench

Thick rods require immense force to start the drawing process. The **Draw Bench** uses mechanical advantage to pull the metal.

A clamp (the "dog") bites the tapered end of the wire protruding from the die. The clamp is attached to a heavy chain or strap. A crank or water wheel winches the chain, pulling the wire through the plate with steady, unstoppable force.

Once the wire is thin enough to be flexible, it is wound onto a capstan drum (the "block"). The rotation of the drum provides the pulling tension, allowing for continuous lengths of wire kilometers long.

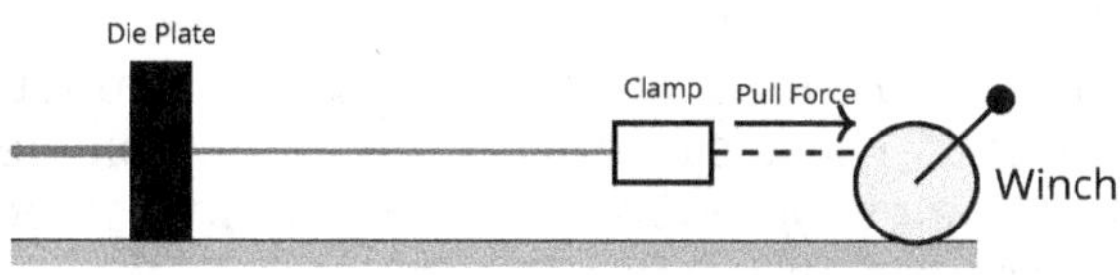

Figure 4: Mechanical advantage applies sufficient tension for heavy gauges.

Summary

Rolling and drawing transform raw ingots into standardized industrial stock. These processes provide the sheet metal for boilers, the rails for trains, and the wire for fences and communications, forming the nervous system and skeleton of the new world.

End of Discovery #019

DISCOVERY #020: POTASH

Category: Chemistry / Materials **Status:** Essential

Soluble Salts from Fire

Potash is a mixture of potassium salts, primarily **Potassium Carbonate** (K_2CO_3). It is obtained by leaching the ashes of hardwood trees. This chemical serves as a critical alkali for the manufacture of glass, soft soaps, and textile bleaches.

During the combustion of wood, organic potassium compounds transform into stable carbonates. These carbonates are highly soluble in water. This property allows for their separation from the insoluble carbon and mineral bulk remaining in the ash after a fire.

The Chemical Cycle

The production process follows a linear extraction path. Raw hardwood is reduced to ash. Water acts as the solvent to remove the salts. Evaporation then recovers the solid chemical.

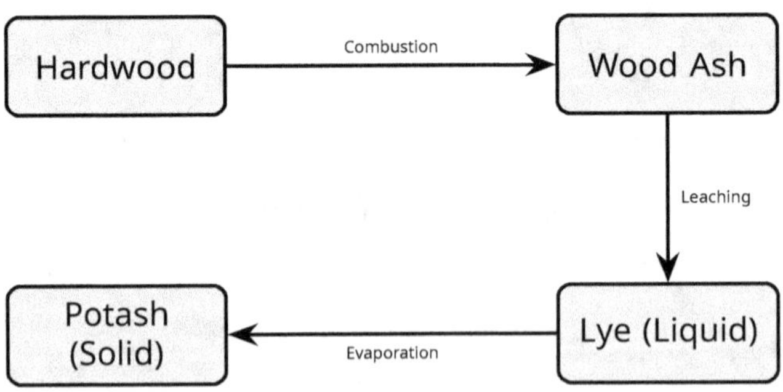

Figure 1: The transformation from organic matter to industrial alkali.

Gravity Separation

The leaching vat is the primary tool for chemical extraction. It uses a layered filtration bed to produce a clear, concentrated alkaline solution known as lye.

The structure is a tapered wooden vessel with a drainage outlet at the base. The layers are packed in a specific order to prevent the fine ash from clogging the outlet or contaminating the liquid solution.

Filter Bed Layers:

- **Ash Body:** Tightly packed hardwood ashes occupy the majority of the volume.
- **Straw Layer:** Acts as a mechanical filter to trap ash particles.
- **Gravel/Stone Layer:** Provides structural support and drainage for the straw.

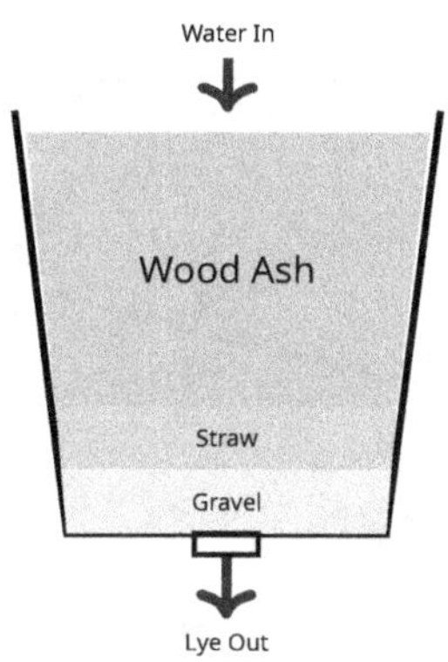

Figure 2: The gravity-fed leaching vat optimizes chemical dissolution.

Concentration Test

The density of the solution indicates the potassium carbonate concentration. A fresh egg or a potato is used as a hydrometer. If the object floats, the lye is sufficiently concentrated for industrial use.

From Black Salts to Pearlash

The initial recovery of potash involves boiling the raw lye in large iron pots until the water evaporates. The residue is a dark mass known as **Black Salts**. This coloration results from residual organic tars and unburnt carbon.

To produce high-purity **Pearlash**, the black salts undergo calcination. They are placed in a reverberatory furnace and subjected to intense heat. The high temperature oxidizes the organic impurities, leaving behind a white, crystalline powder of refined potassium carbonate.

Solubility and Recovery

The efficiency of the recovery process depends on the high solubility of K_2CO_3 in water. This solubility increases significantly with temperature, allowing for the creation of saturated solutions that yield high volumes of salt upon evaporation.

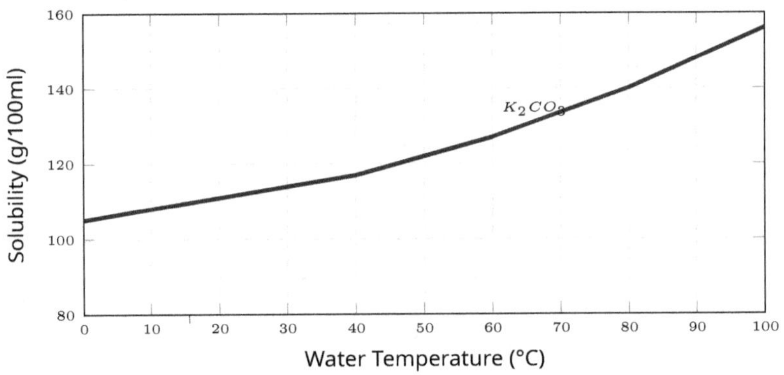

Figure 3: Solubility curve demonstrating extraction efficiency.

Handling Precautions

Concentrated lye is a strong caustic base. It dissolves organic tissues and causes chemical burns. Wood, iron, and ceramics are suitable materials for lye equipment. Aluminum and lead react with the alkali and must be avoided.

The Chemistry of Cleansing

The primary industrial application of potash is the process of **Saponification**. Heating potash lye with animal fats (tallow) or vegetable oils transforms the lipids into glycerol and potassium salts of fatty acids.

Sodium carbonate produces hard bar soap. Potassium-based lye produces **Soft Soap**, which remains a semi-liquid paste. This product effectively removes grease and oils from surfaces and fibers.

Industrial Uses:

- **Glassmaking:** Acts as a flux to lower the melting point of silica.
- **Textiles:** Used for "scouring" raw wool to remove natural lanolin.
- **Bleaching:** Prepares fibers for dyeing by removing natural pigments.

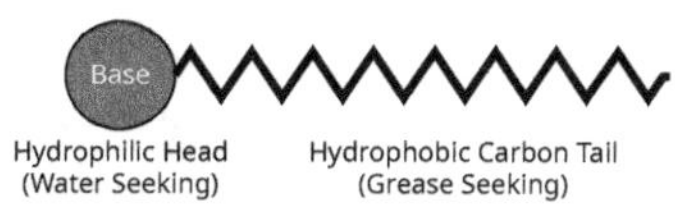

$$\text{Fat} + \text{Potash Lye} \rightarrow \text{Soft Soap} + \text{Glycerol}$$

Figure 4: The molecular structure of soap enables the emulsification of oils.

Summary

Potash extraction represents the beginning of industrial chemistry. By utilizing the waste products of wood combustion, we produce a versatile alkali that facilitates hygiene, construction, and textile manufacturing.

End of Discovery #020

DISCOVERY #021: SOAP

Category: Chemistry / Hygiene **Status:** Essential

The Chemistry of Saponification

Soap is a surfactant produced through the chemical hydrolysis of triglycerides. This reaction occurs when a lipid (fat or oil) reacts with a strong alkaline base, such as potassium hydroxide (from wood ash lye) or sodium hydroxide.

The process breaks the ester bonds of the triglyceride molecule. One molecule of fat reacts with three molecules of the alkaline base to produce one molecule of glycerol and three molecules of soap. The resulting substance is a salt of a fatty acid.

Molecular Stoichiometry

The efficiency of the reaction depends on the correct ratio of reactants. Excess lye results in a caustic product that irritates the skin. Excess fat results in a greasy soap that lacks effective cleaning power.

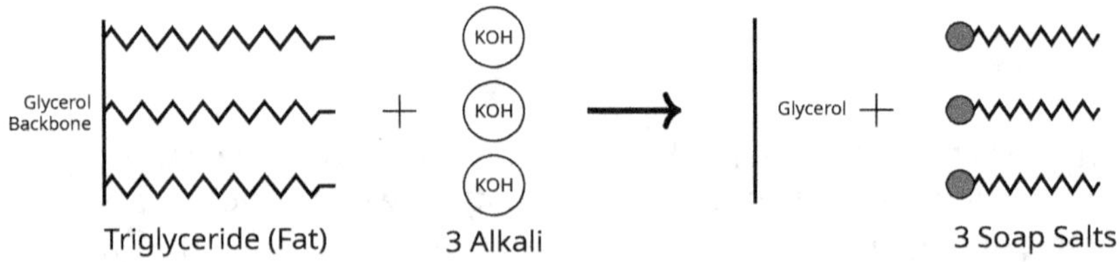

Figure 1: The alkaline base detaches fatty acid chains from the glycerol backbone.

The Boiling Method

The hot process utilizes thermal energy to accelerate the saponification reaction. Reactants are mixed in a large iron cauldron and heated to a steady boil. Constant agitation ensures that the immiscible fat and water-based lye maintain contact.

Process Stages:

1. **Melting:** Solid fats are liquefied.
2. **Emulsification:** Lye is added gradually while stirring.
3. **Saponification:** The mixture thickens into a "trace" state, resembling pudding.
4. **Graining Out:** Salt is added to the mixture. This increases the ionic strength of the liquid, causing the soap to precipitate and float to the surface.

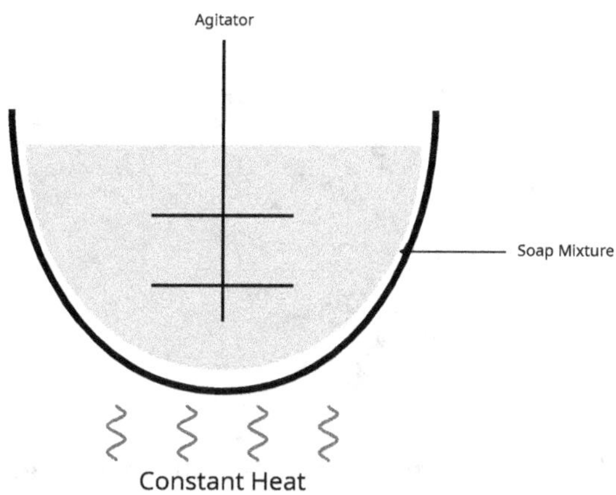

Figure 2: Continuous stirring and heat ensure a complete chemical conversion.

Refining the Product

After graining, the glycerol and excess water settle at the bottom. The top layer of soap is skimmed off, poured into molds, and left to harden.

Surfactant Action

Soap functions as an emulsifier through its dual-nature molecular structure. Each soap molecule possesses a **hydrophilic head** and a **hydrophobic tail**.

The head is polar and carries an ionic charge, allowing it to bond with water molecules. The tail is a long hydrocarbon chain that is non-polar and attaches to oils, grease, and fats. This configuration allows the soap to bridge the gap between substances that do not naturally mix.

Micelle Formation

When soap is introduced to water containing grease, the molecules organize into spherical structures called **micelles**. The hydrophobic tails point inward, trapping the oil at the center. The hydrophilic heads point outward, maintaining a bond with the surrounding water. This enables the grease to be lifted from a surface and washed away.

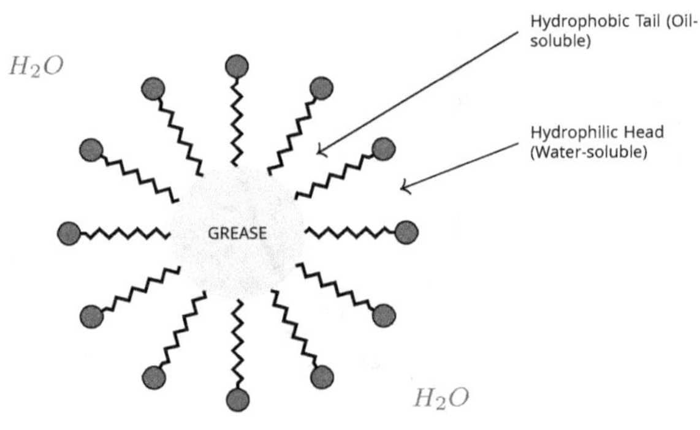

Figure 3: Micelles isolate grease particles within a water-soluble shell.

Potassium vs. Sodium Soaps

The type of alkaline base determines the physical consistency of the final soap. Potassium hydroxide (derived from hardwood ash lye) produces a semi-liquid or soft soap. Sodium hydroxide produces a solid, hard bar soap.

The addition of common salt ($NaCl$) during the boiling of potassium soap can facilitate a partial ion exchange, resulting in a firmer consistency.

Curing and pH

Freshly made soap requires a curing period of several weeks. During this time, the water content evaporates, increasing the hardness and longevity of the bar. Furthermore, the saponification reaction completes any residual chemical processes, ensuring the pH level drops to a range safe for human skin (pH 8-10).

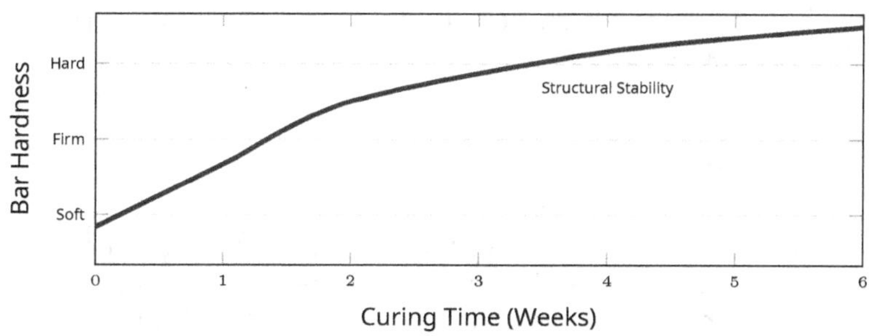

Figure 4: The curing process improves the durability and safety of the product.

Summary

Soap production transforms raw waste fats and wood ash into a fundamental tool for disease prevention. By mastering the saponification reaction, a civilization establishes the chemical basis for public health, significantly reducing the transmission of pathogens.

End of Discovery #021

DISCOVERY #022: LEATHER TANNING

Category: Materials / Biology **Status:** Essential

The Collagen Lattice

Animal skins are composed primarily of **collagen**, a structural protein organized in a complex, three-dimensional fibrous lattice. In its raw state, this material is subject to rapid bacterial decomposition when moist and becomes hard and translucent when dry.

Tanning is the chemical process of stabilizing this protein structure. It involves the permanent cross-linking of collagen fibers, rendering the material resistant to decay, heat, and moisture. This transformation converts a perishable byproduct into leather, a durable industrial material essential for belts, valves, footwear, and armor.

Hide Structure

The tanning process focuses on the **corium**, the thickest layer of the skin located beneath the epidermis. The removal of hair and non-collagenous proteins through liming exposes the fibrous network for chemical treatment.

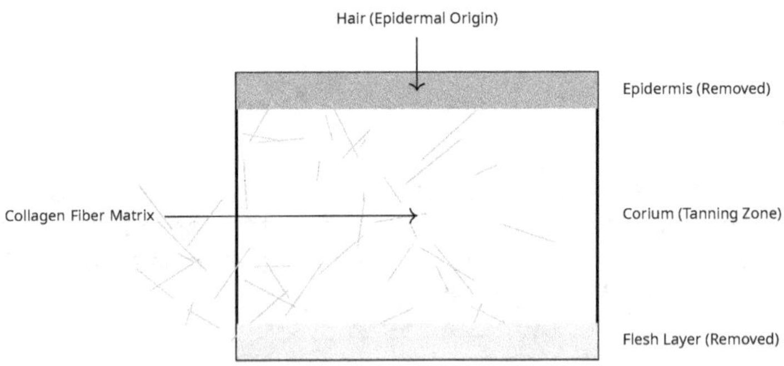

Figure 1: The removal of surface layers exposes the corium for chemical stabilization.

Polyphenolic Agents

Vegetable tanning utilizes **tannins**, a class of polyphenolic compounds found in the bark, wood, and leaves of specific trees, such as oak, chestnut, and hemlock. Tannins possess the chemical ability to bind strongly to the amino groups in collagen.

The concentration of tannins varies by species and maturity. Extraction involves grinding the bark and leaching it with warm water to create a concentrated infusion.

The Binding Mechanism

Tannin molecules displace water from the collagen fibers and form hydrogen bonds between the protein chains. This displacement prevents the fibers from adhering to one another upon drying, maintaining the flexibility and structural integrity of the skin.

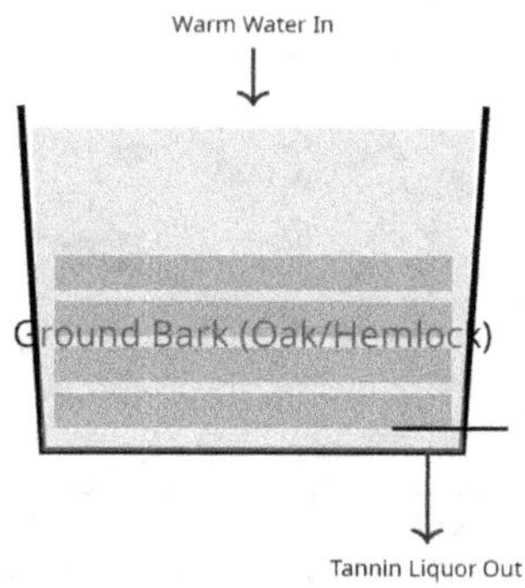

Figure 2: Leaching ground bark produces the polyphenolic agents required for tanning.

Sequential Immersion

Effective tanning requires the gradual penetration of tannins into the thickest parts of the hide. Rapid exposure to high concentrations of tannin causes "dead tanning," where the surface fibers seal prematurely, preventing the core from being treated.

The pit system utilizes a series of vats with increasing liquor concentrations. Hides progress through these pits over several months.

Processing Stages:

1. **Suspenders:** Hides hang in weak, used liquors to begin primary fiber separation.
2. **Handlers:** Hides move to stronger solutions and undergo regular agitation.
3. **Layers:** Hides lie flat in pits with ground bark between each layer, submerged in the strongest tannin infusion for final saturation.

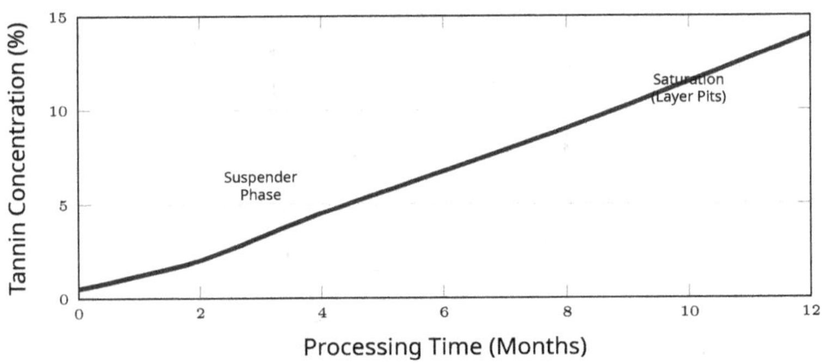

Figure 3: Gradual increases in concentration ensure total penetration of the hide.

Lubrication of Fibers

Tanned leather is initially stiff and brittle. The removal of water leaves voids between the cross-linked collagen fibers. Fibers that remain unlubricated grind against each other, leading to mechanical failure and cracking.

Fat Liquoring is the application of oils and fats (such as tallow, neatsfoot oil, or fish oils) to the damp leather. These lipids coat the individual fibers, providing internal lubrication. This step determines the final softness, water resistance, and tensile strength of the leather.

Mechanical Finishing

Once lubricated, the leather is "staked" or worked over a blunt metal blade. This physical manipulation separates fibers that have adhered during drying, resulting in a flexible, high-performance material.

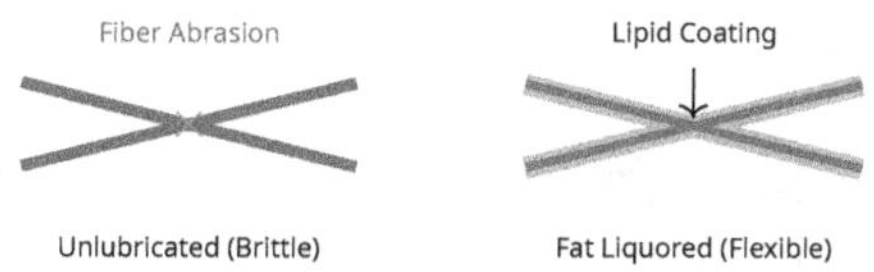

Figure 4: Lipid application prevents fiber cohesion and ensures mechanical longevity.

Summary

Leather tanning converts organic waste into a versatile industrial polymer. By utilizing vegetable tannins and lipid lubrication, a civilization produces a material capable of withstanding extreme mechanical stress and environmental exposure.

End of Discovery #022

DISCOVERY #023: ALCOHOL DISTILLATION

Category: Chemistry / Energy **Status:** Essential

Differential Boiling Points

Distillation is the process of separating components from a liquid mixture through selective evaporation and condensation. This technique relies on the unique boiling points of different substances. Ethanol (C_2H_5OH) possesses a boiling point of $78.3°$C at standard atmospheric pressure. Water boils at $100°$C.

When a fermented solution is heated to a temperature between these two points, the ethanol transitions into a vapor phase more rapidly than the water. Capturing and cooling this vapor yields a liquid with a significantly higher alcohol concentration.

Vapor-Liquid Equilibrium

The concentration of ethanol in the vapor is determined by the Vapor-Liquid Equilibrium (VLE). In a simple distillation cycle, the vapor phase is richer in the more volatile component (ethanol) than the boiling liquid phase.

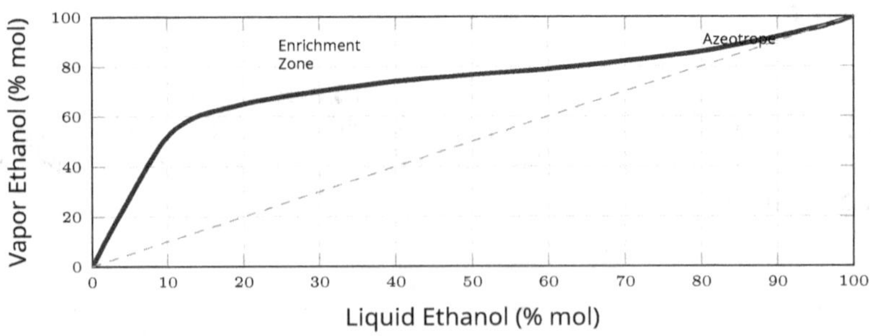

Figure 1: The equilibrium curve illustrates the alcohol enrichment during evaporation.

The Pot Still

The primary apparatus for distillation is the **Pot Still**. It consists of four major functional components designed to manage the transition of states.

- **Cucurbit (Pot):** The vessel containing the fermented "wash." Heat is applied here to initiate evaporation.
- **Alembic (Head):** The dome-shaped cap where vapors collect. Its shape prevents liquid splashing from entering the condenser.
- **Lyne Arm:** A slanted pipe that conducts vapor from the head to the cooling zone.
- **Condenser (Worm):** A coiled tube submerged in cold water. The vapor releases thermal energy to the water and returns to a liquid state.

Thermal Management

Maintaining a steady temperature is critical. Overheating causes excessive water evaporation, diluting the distillate. A "bain-marie" ensures uniform heat distribution and prevents scorching of organic solids.

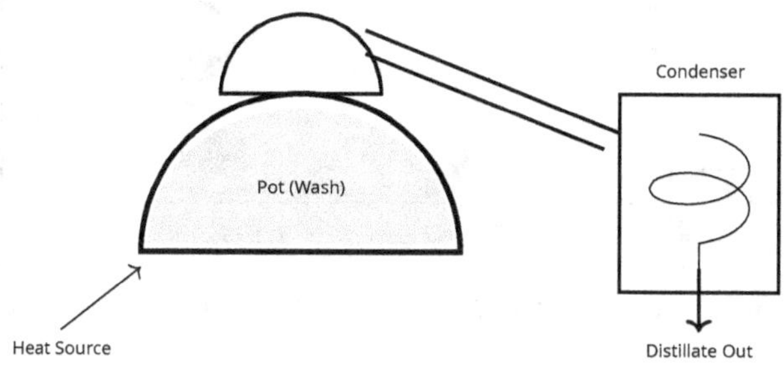

Figure 2: The modular assembly allows for controlled evaporation and rapid cooling.

Heads, Hearts, and Tails

The distillation output is collected in three distinct fractions. Each fraction contains different chemical compounds based on their volatility.

1. Heads (Foreshots): The first liquid to emerge. It contains methanol and acetaldehydes. This fraction is toxic and is set aside for use as a solvent or fuel.

2. Hearts: The middle fraction containing high-purity ethanol. It is clear and represents the primary goal of the distillation process.

3. Tails (Feints): The final fraction. It contains heavier alcohols and esters. These are collected and added to the next batch for re-distillation.

Concentration Cycles

A single distillation of a 10% ABV wash typically yields a 30-40% ABV distillate. Re-distilling this product (double distillation) increases the concentration to 70-80%.

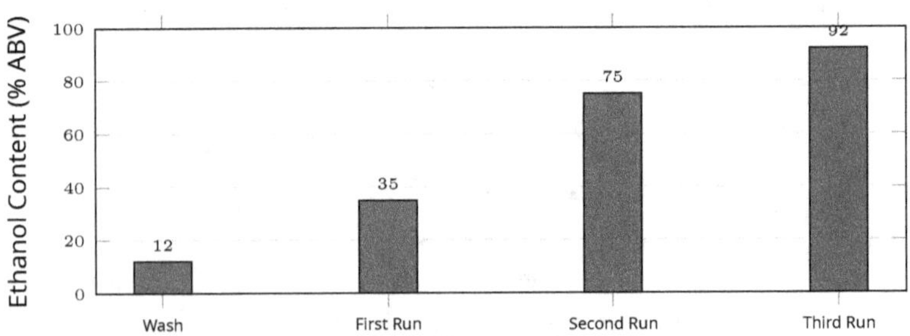

Figure 3: Sequential distillation stages incrementally increase chemical purity.

Antisepsis and Solvent Action

At 70% concentration, ethanol acts as a powerful broad-spectrum antimicrobial agent. It denatures proteins and dissolves lipid membranes of bacteria and viruses, making it essential for medical sterilization. As a solvent, it dissolves resins and oils insoluble in water.

Liquid Fuel

Ethanol is a high-energy density fuel. It burns cleanly with a blue flame, producing carbon dioxide and water:

$$C_2H_5OH + 3O_2 \rightarrow 2CO_2 + 3H_2O + \text{Heat}$$

In internal combustion engines, it provides a renewable energy source.

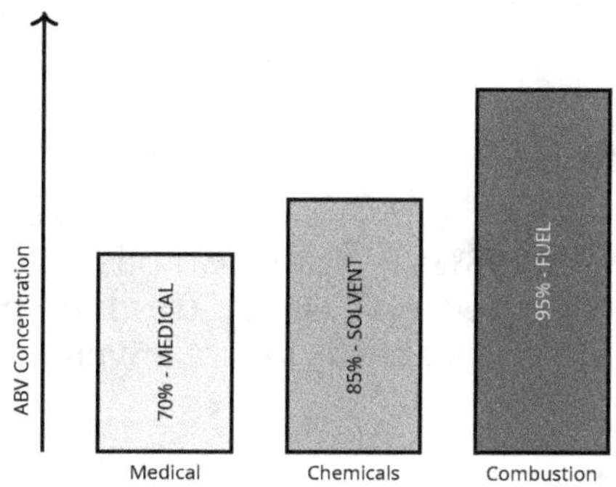

Figure 4: Industrial utility increases with ethanol concentration.

Summary

Alcohol distillation transforms agricultural surplus into a critical resource. By isolating ethanol, a civilization gains effective medicine, advanced chemical processing, and renewable energy.

End of Discovery #023

DISCOVERY #024: SULFURIC ACID

Category: Chemistry / Industry **Status:** Essential

The King of Chemicals

Sulfuric acid (H_2SO_4) is a dense, colorless, and oily liquid. Its chemical versatility makes it the primary reagent for heavy industry. It acts as a powerful mineral acid, a strong oxidizing agent, and a unique dehydrating agent.

The molecular structure consists of a central sulfur atom bonded to four oxygen atoms. Two of these oxygen atoms are part of hydroxyl groups (—OH), which release hydrogen ions in aqueous solutions to provide acidic properties.

Density and State

In its concentrated form (98%), sulfuric acid has a specific gravity of 1.84, making it nearly twice as dense as water. This high density allows for the storage of immense chemical energy in a relatively small volume.

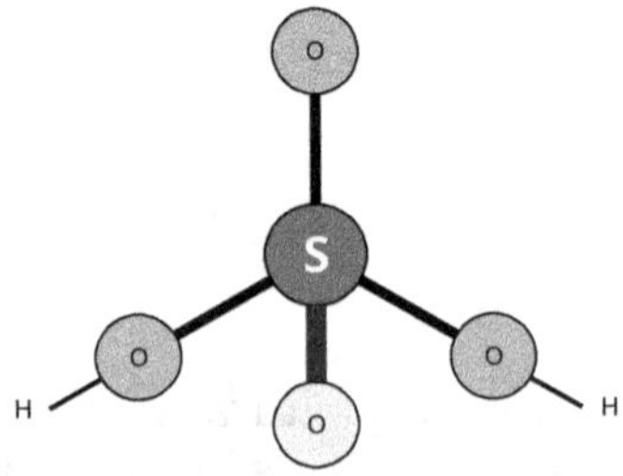

Tetrahedral Geometry of H_2SO_4

Figure 1: The arrangement of oxygen atoms around sulfur determines its high reactivity.

The Vitriol Process

Early production relies on the thermal decomposition of mineral sulfates, historically known as **vitriols**. Heating Green Vitriol (Iron II Sulfate, $FeSO_4 \cdot 7H_2O$) in a ceramic retort releases sulfur trioxide gas, which is subsequently dissolved in water to form the acid.

The Bell Process

The transition to industrial quantities utilizes the combustion of elemental sulfur mixed with saltpeter (Potassium Nitrate).

1. **Oxidation:** Sulfur is burned to create Sulfur Dioxide (SO_2).
2. **Catalysis:** Saltpeter provides nitrogen oxides that catalyze the oxidation of SO_2 into Sulfur Trioxide (SO_3).
3. **Hydration:** The SO_3 gas reacts with water vapor to form a mist of H_2SO_4.

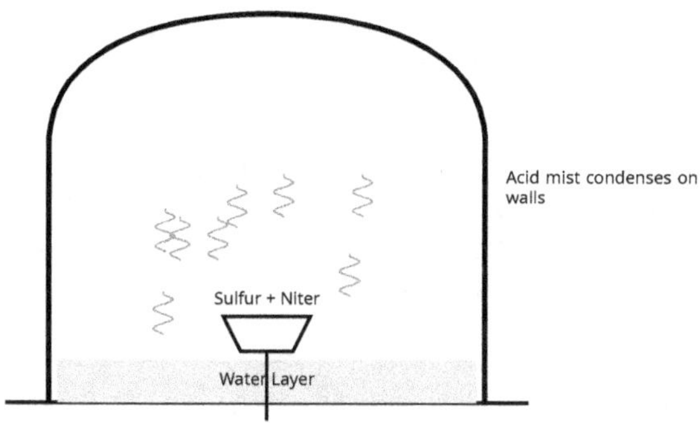

Figure 2: Early industrial production using the sulfur-niter combustion cycle.

Dehydration and Affinity for Water

Sulfuric acid possesses a high affinity for water. When concentrated acid is added to organic materials (such as wood, sugar, or paper), it removes hydrogen and oxygen atoms in the ratio of water ($2:1$). This leaves behind a residue of pure carbon.

Heat of Dilution

The reaction between sulfuric acid and water is extremely exothermic. Significant thermal energy is released during mixing. **Safety Protocol:** Always add acid to water. Adding water to concentrated acid causes the top layer to boil instantly, resulting in dangerous acidic spray.

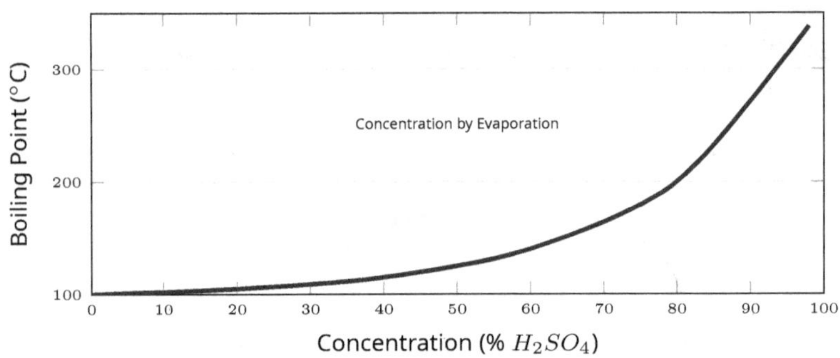

Figure 3: The boiling point increases sharply as the acid reaches high concentration.

Industrial Utility

The consumption of sulfuric acid serves as a direct indicator of a civilization's industrial capacity. Its primary applications span across every sector of manufacturing.

- **Fertilizers:** Used to treat phosphate rock to produce "superphosphate," making phosphorus available for plant uptake.
- **Metallurgy:** Removes oxides from the surface of iron and steel (pickling) before further processing.
- **Chemical Synthesis:** Acts as a catalyst in the production of explosives (Nitroglycerin), dyes, and drugs.
- **Petroleum Refining:** Catalyzes the alkylation process and removes impurities from fuel.

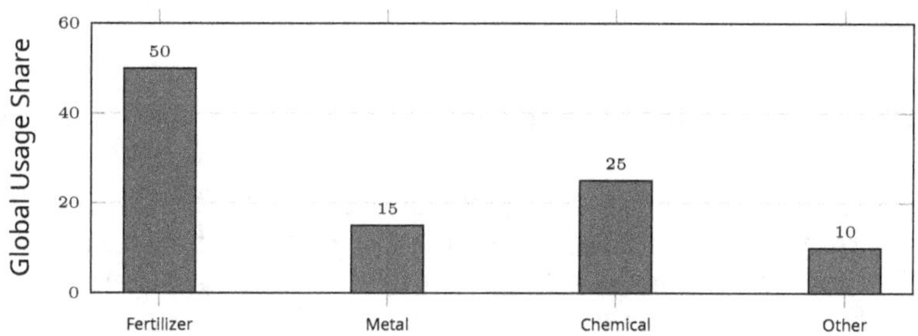

Figure 4: The majority of sulfuric acid production supports agricultural productivity.

Summary

Sulfuric acid is the fundamental liquid of industrial chemistry. It enables the mass production of fertilizers, metals, and complex chemicals, providing the necessary reagents for a self-sustaining technological society.

End of Discovery #024

DISCOVERY #025: BLACK POWDER

Category: Chemistry / Civil Engineering **Status:** Essential

The Ternary Mixture

Black powder is a mechanical propellant composed of three distinct ingredients: **Saltpeter** (Potassium Nitrate), **Charcoal**, and **Sulfur**. The standard industrial ratio for mining and engineering is 75% saltpeter, 15% charcoal, and 10% sulfur.

Each component serves a specific thermodynamic function. Saltpeter acts as the oxidizer, providing the oxygen necessary for combustion in confined spaces. Charcoal provides the carbon fuel. Sulfur acts as a stabilizer and lowers the ignition temperature, accelerating the rate of the chemical reaction.

The Deflagration Reaction

Upon ignition, the mixture undergoes rapid deflagration. The reaction converts solid matter into high-temperature gases (Nitrogen, Carbon Dioxide) and solid residue (Potassium Carbonate). The rapid expansion of these gases generates the pressure required for mechanical work.

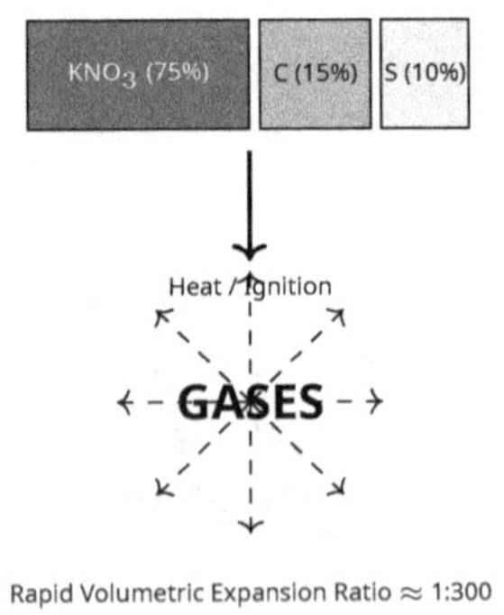

Figure 1: The transformation of solid reactants into expanding gaseous products.

Incorporation and Milling

Mechanical mixing is insufficient for consistent performance. The ingredients must be incorporated through intensive milling. A **Stamp Mill** uses heavy pressure to crush the components together at the microscopic level.

Moisture (water or alcohol) is added during the milling process to prevent accidental ignition and to ensure the saltpeter impregnates the pores of the charcoal. This intimate contact is necessary for the simultaneous reaction of all particles.

Pressing the Cake

After milling, the "mill cake" is subjected to high pressure using a screw or hydraulic press. This increases the density of the powder, which directly controls the burn rate. High-density powder burns predictably, making it suitable for blasting hard rock in mining operations.

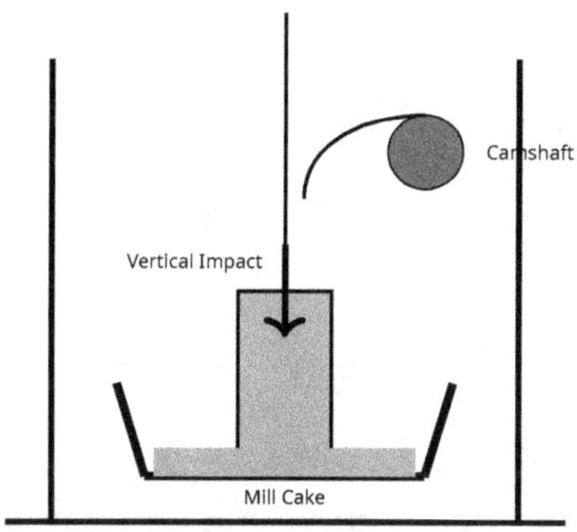

Figure 2: Stamp mill mechanism for the microscopic incorporation of ingredients.

The Corning Process

Pressed powder is broken into smaller pieces through a process called **Corning**. These pieces are passed through a series of sieves to classify them by size. Granulated powder allows air gaps between the grains, which enables the flame to spread instantly through the entire charge.

Grain Size Classification

Grain size determines the surface area available for combustion. Larger grains (1.5mm - 3.0mm) burn slowly and are used for heavy blasting in soft rock. Smaller grains (0.5mm - 1.0mm) burn faster and are used for hard rock fragmentation.

- **Cg (Coarse):** Standard for mining and canal excavation.
- **Fg (Fine):** Standard for precision blasting and small-bore drilling.
- **Dust:** Material for re-milling or fuse production.

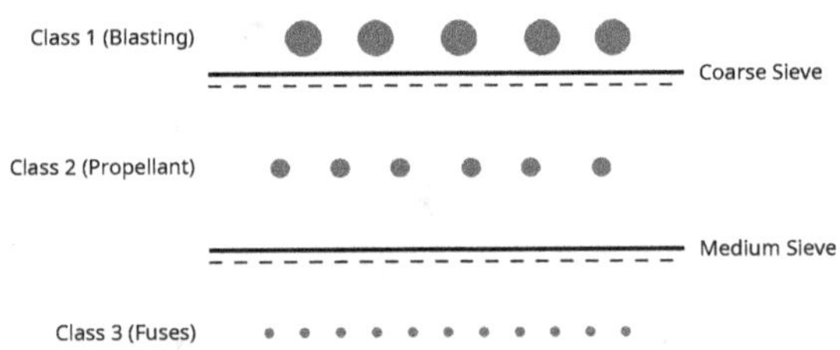

Figure 3: Sieving process for standardizing burn rates through granulation.

Borehole Blasting

In mining and civil engineering, black powder is placed in deep boreholes drilled into the rock face. The charge is confined with compacted clay or sand to ensure the pressure is directed against the rock walls.

Black powder is a "heaving" explosive. It generates a sustained push that creates large fractures and moves volumes of material. This allows for the extraction of stone blocks without pulverizing the material.

The Industrial Impact

The use of black powder reduces manual labor for infrastructure. It is fundamental for:

- **Canal Construction:** Rapid excavation of shipping routes.
- **Tunnelling:** Penetrating mountains for transportation.
- **Mining:** Accessing deep mineral veins previously unreachable.

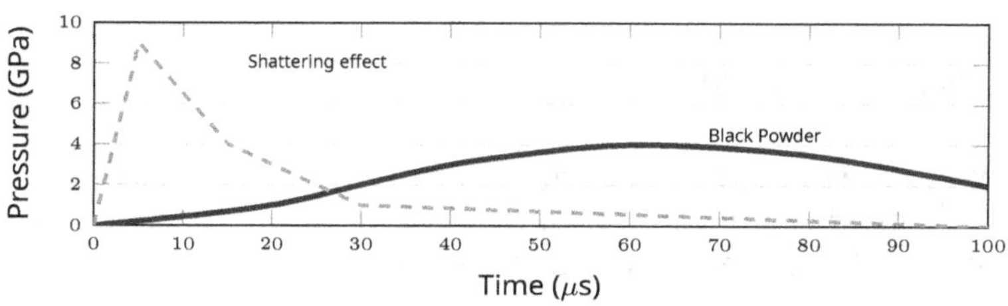

Figure 4: The pressure-time profile showing the heaving action.

Summary

Black powder is the first chemical tool to harness energy for construction. Its predictable behavior and heaving action make it essential for earthmoving and mineral extraction.

End of Discovery #025

DISCOVERY #026: RUBBER VULCANIZATION

Category: Materials / Chemistry **Status:** Essential

The Molecular Lattice

Natural rubber is an elastomer composed of long polymer chains of **cis-polyisoprene** $(C_5H_8)_n$. These molecules exist as disordered, tangled coils. A mechanical force applied to the material causes the chains to straighten. Upon release, the internal entropy of the system pulls the chains back to their original coiled state, providing the material with its characteristic elasticity.

Raw rubber exhibits high temperature sensitivity. It enters a liquid state in heat and becomes brittle in cold. This thermoplastic behavior limits its industrial utility to low-stress applications and temperate environments.

Elastomeric Geometry

The functionality of rubber depends on the freedom of movement within the polymer chains. In the raw state, these chains slide past one another under sustained tension, resulting in permanent deformation or creep.

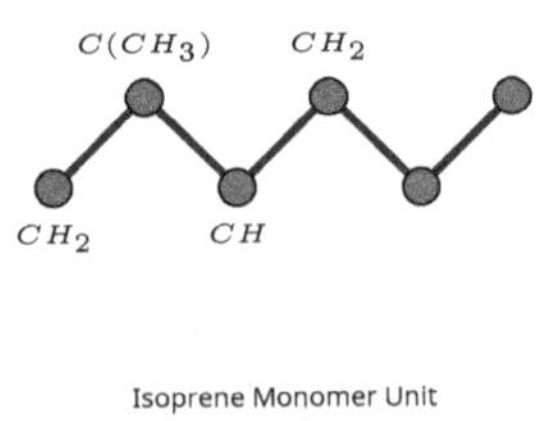

Figure 1: The repeating units of polyisoprene form the basis of natural elastic polymers.

The Vulcanization Reaction

Vulcanization is the chemical process of introducing **sulfur bridges** between the long polyisoprene chains. Heating the raw rubber with elemental sulfur (S) and accelerators initiates a reaction at the double-bond sites of the polymer.

These cross-links act as chemical anchors. They prevent the individual chains from sliding past each other while allowing them to stretch and recoil. This transformation creates a stable, three-dimensional molecular network that retains its shape under stress and varying temperatures.

Cross-Link Density

The physical properties of the vulcanized rubber are determined by the concentration of sulfur. Low sulfur content (1–3%) produces soft, highly elastic rubber suitable for tires and seals. High sulfur content (up to 30%) produces a rigid, hard material known as ebonite, used for structural components and insulators.

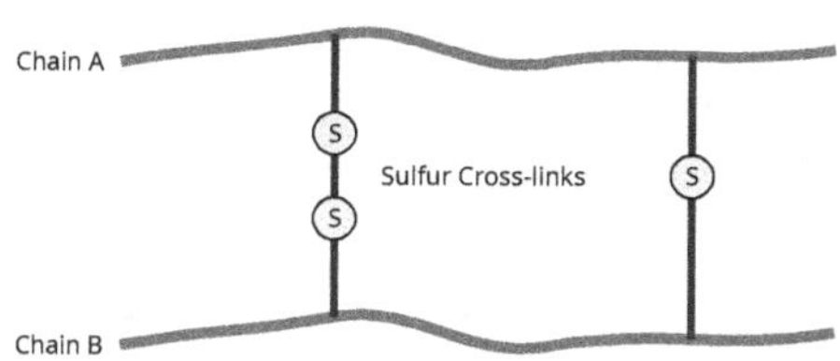

Figure 2: Sulfur atoms form covalent bonds that lock polymer chains into a fixed lattice.

Mechanical Resilience

Vulcanized rubber maintains its mechanical properties across a wide thermal range. The cross-links prevent the transition into a viscous liquid state at high temperatures and resist the alignment of chains that causes crystallization and brittleness at low temperatures.

This stability is critical for industrial applications where the material is subjected to frictional heat or environmental exposure. The modulus of elasticity remains relatively constant, providing predictable performance for engineering components.

Tension and Recovery

A vulcanized elastomer returns to its precise original dimensions after being stretched to several times its original length. This ability to absorb and release energy without permanent deformation enables the creation of high-efficiency vibration dampers and springs.

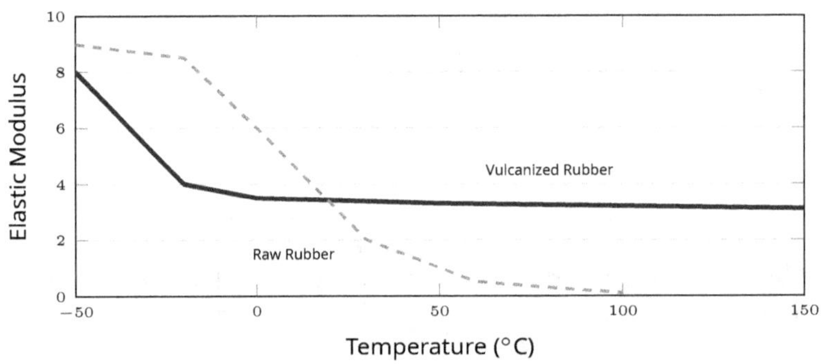

Figure 3: Vulcanization provides stable elastic behavior across industrial temperature ranges.

Industrial Sealing Technology

The primary utility of vulcanized rubber in engineering is the creation of **hermetic seals** and **gaskets**.conforming to surface irregularities under pressure, rubber prevents the leakage of gases and liquids in mechanical assemblies.

Gasket Mechanics: Compression of a rubber gasket between two rigid flanges exerts an outward force against the sealing surfaces. This contact pressure must exceed the internal pressure of the fluid being contained. The material's resilience ensures the seal remains intact during vibrations or thermal expansion.

Waterproofing and Insulation

Vulcanized rubber is impermeable to water and highly resistant to electrical conductivity. It serves as the standard material for waterproofing protective garments and providing flexible insulation for electrical systems.

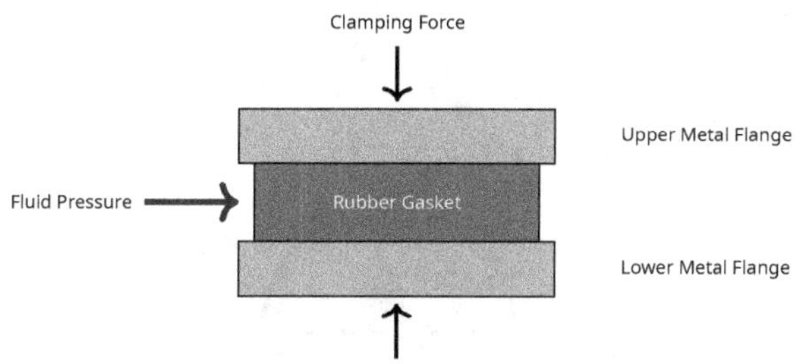

Figure 4: Mechanical compression of an elastomer creates a high-pressure hermetic seal.

Summary

Vulcanization converts a temperature-sensitive organic extract into a high-performance industrial polymer. By establishing chemical cross-links, a civilization gains the capability to contain high-pressure fluids and protect delicate systems from environmental exposure.

End of Discovery #026

DISCOVERY #027: WHEEL AND AXLE

Category: Mechanics / Transport **Status:** Essential

The Rolling Principle

Transporting heavy loads across terrestrial surfaces requires overcoming the force of friction. Sliding an object generates significant kinetic friction, as the entire contact surface of the load interacts with the ground. The wheel converts sliding friction into rolling resistance.

In a wheel and axle system, the only sliding contact occurs at the interface between the axle and the hub. Because the axle has a much smaller diameter than the wheel, the distance traveled by the friction-generating surface is drastically reduced. This geometric arrangement allows for the movement of massive weights with minimal energy expenditure.

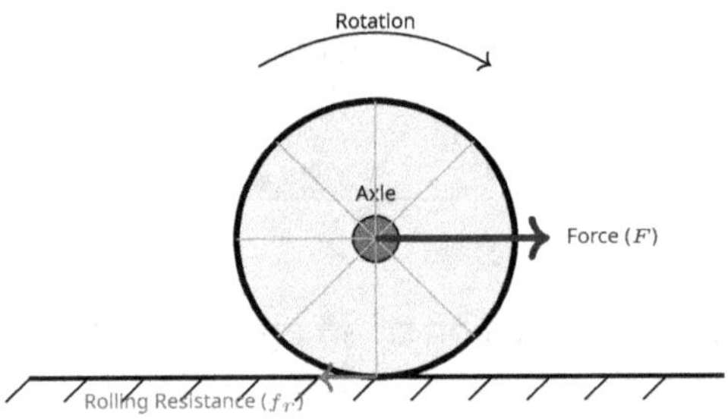

Figure 1: The wheel concentrates frictional interaction to the small surface area of the axle.

Spoked Construction

The structural efficiency of a wheel is determined by the distribution of radial stress. A solid wooden wheel provides high durability but adds excessive dead weight to the vehicle. The spoked wheel utilizes a rigid rim connected to a central hub by tensile or compressive members.

This design reduces the rotational inertia of the system, allowing for faster acceleration and better maneuverability. The rim is often reinforced with a metallic tire (iron or steel) to prevent abrasion of the wood and to provide a "shrink-fit" that compresses the assembly into a single, high-tension unit.

Axle Integrity

The axle must withstand the shear stress of the load while maintaining a smooth surface for rotation. Hardwood axles require regular lubrication with animal fats or mineral oils to prevent thermal failure caused by friction.

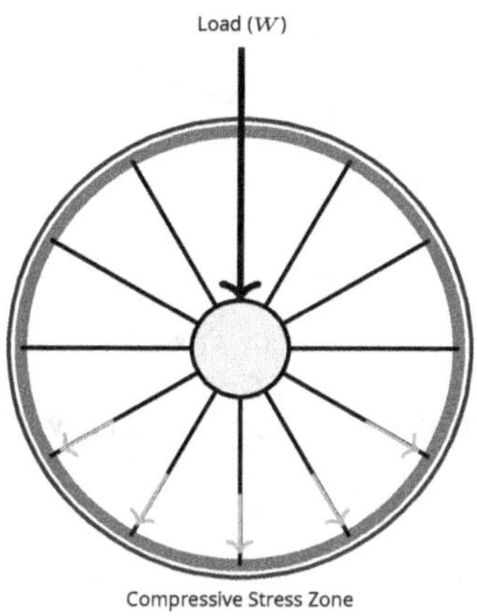

Figure 2: Spokes distribute the vertical load across the lower arc of the rim.

The Wheel as a Lever

The wheel and axle function as a continuous lever system. The radius of the wheel acts as the effort arm, while the radius of the axle acts as the resistance arm. This relationship creates a mechanical advantage (MA) that enables the application of high torque.

$$MA = \frac{Radius_{wheel}}{Radius_{axle}}$$

In applications such as a windlass or a steering wheel, a large input radius allows a human operator to generate the force necessary to lift heavy buckets of water or redirect the momentum of a sailing vessel. Conversely, applying power to the axle results in high rotational speed at the wheel's rim, which is essential for rapid land transport.

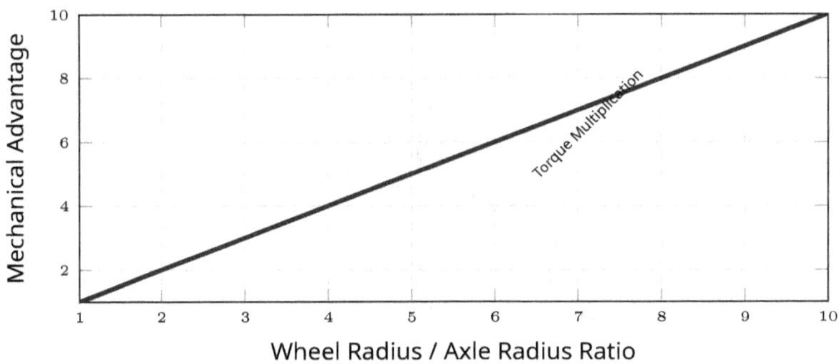

Figure 3: Force multiplication increases linearly with the radius of the wheel.

Transportation Efficiency

The implementation of the wheel fundamentally alters the logistics of a civilization. Efficiency is measured in ton-kilometers—the amount of weight that can be moved per unit of energy. A single draft animal pulling a wheeled cart can move five to ten times the weight possible with a sledge or pack saddle.

This surplus capacity enables the transport of heavy construction materials (stone, timber) and bulk food supplies over long distances. The resulting ability to centralize resources facilitates the growth of urban centers and specialized industrial zones.

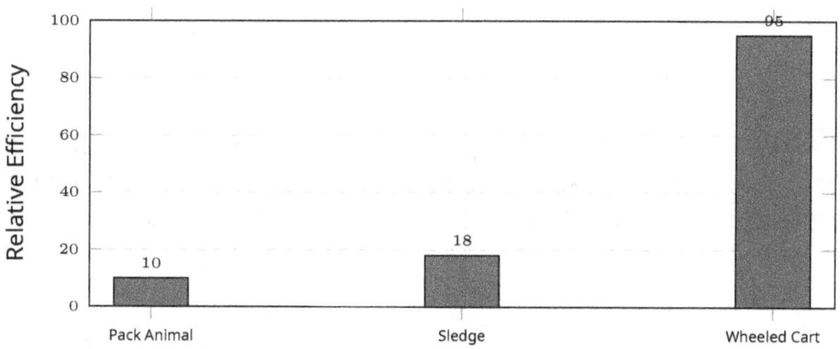

Figure 4: The wheeled cart provides the highest efficiency for terrestrial freight.

Summary

The wheel and axle convert linear force into rotational motion, significantly reducing the energy required for transport. This discovery establishes the foundation for land-based commerce and large-scale infrastructure development.

End of Discovery #027

DISCOVERY #028: LIFTING MACHINES

Category: Physics / Construction **Status:** Essential

The Law of the Lever

The lever is a rigid beam pivoting on a fixed point called a **fulcrum**. It operates on the principle of moments, where the force applied at one end is multiplied based on its distance from the pivot. The equilibrium of a lever is defined by the formula:

$$F_1 \cdot d_1 = F_2 \cdot d_2$$

A small force applied over a large distance can lift a heavy weight over a short distance. This trade-off between force and displacement allows a single operator to manipulate objects that far exceed their biological strength.

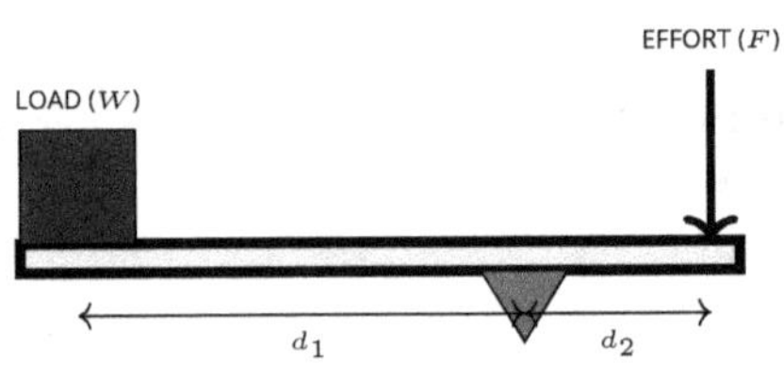

Figure 1: The mechanical advantage is determined by the ratio of the lever arms.

Tension Distribution

A pulley is a wheel on an axle designed to support the movement and change of direction of a cable. A single fixed pulley provides no mechanical advantage but alters the direction of force, allowing an operator to use their body weight to lift a load.

A **Block and Tackle** system uses multiple pulleys (fixed and movable) to distribute the load across several segments of the rope. Each additional segment of rope supporting the load reduces the force required by the operator.

$$F = \frac{W}{n}$$

where n is the number of rope segments supporting the movable block.

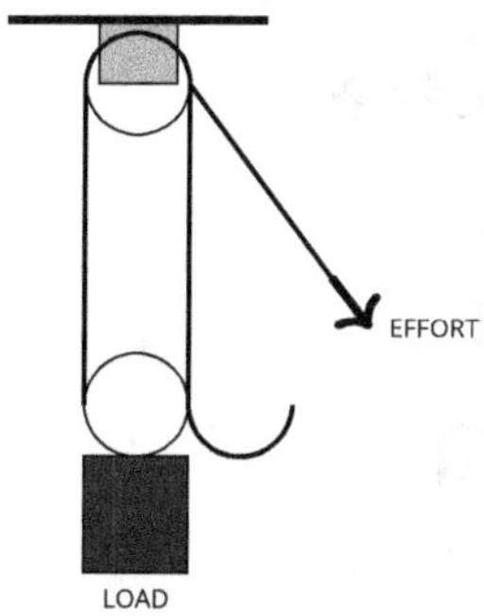

Figure 2: A two-pulley system halves the effort required to lift the load.

Rotational Work

The winch (or windlass) is a device for hauling or lifting that utilizes a horizontal cylinder and a crank. It operates as a continuous lever, where the crank handle acts as the effort arm and the cylinder radius acts as the resistance arm.

The use of a winch allows for the precise, controlled movement of heavy loads over long distances, such as raising blocks of stone in crane assemblies or drawing water from deep wells. A vertical variation, the **capstan**, is used for heavy horizontal hauling, such as moving ships or tensioning large cables.

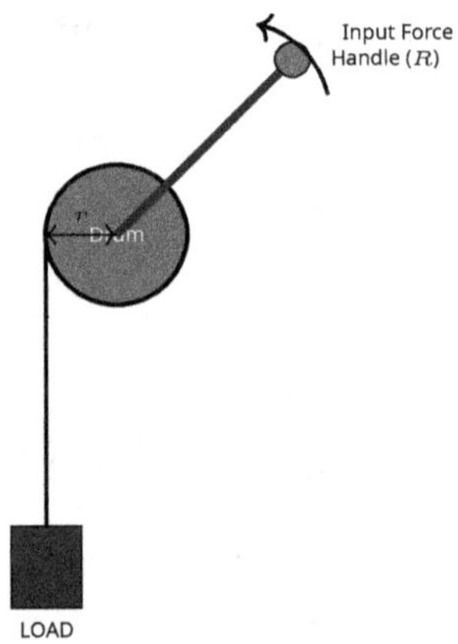

Figure 3: The winch provides high mechanical advantage for vertical lifting.

Architectural Synthesis

Complex construction requires the integration of multiple simple machines. The **Treadwheel Crane** combines the lever (in its frame), the pulley (in its hoist), and the winch (driven by a large wheel). Human operators inside the wheel provide a steady, high-torque input that can lift several tons of masonry.

These machines enable the construction of vertical structures, such as aqueducts, cathedrals, and defensive walls. The ability to lift concentrated loads to significant heights is the prerequisite for advanced architecture.

Machine System	Typical Ratio	Output for 50kg Input
Single Lever (long)	10:1	500 kg
Block and Tackle (4)	4:1	200 kg
Winch with Crank	15:1	750 kg
Treadwheel Crane	40:1	2,000 kg

Table 1: Comparison of mechanical advantages in lifting technologies.

Summary

Lifting machines bypass the biological limits of human muscle. By exchanging distance for force, a civilization can manipulate its environment on a monumental scale, creating the infrastructure of a modern society.

End of Discovery #028

DISCOVERY #029: DOUBLE-ACTION BELLOWS

Category: Mechanics / Metallurgy **Status:** Essential

The Principle of Constant Blast

Smelting and forging operations require high, stable temperatures. Traditional single-action bellows provide air only during the compression stroke, causing temperature fluctuations in the furnace. The double-action bellows utilizes a dual-chamber system to deliver a continuous stream of oxygen to the fire.

The device consists of an intake chamber and a pressurized reservoir. As air is pumped into the lower chamber, a portion of the volume is forced into the upper reservoir. The weight of the upper board maintains a steady pressure, forcing air through the nozzle even during the intake cycle of the primary pump.

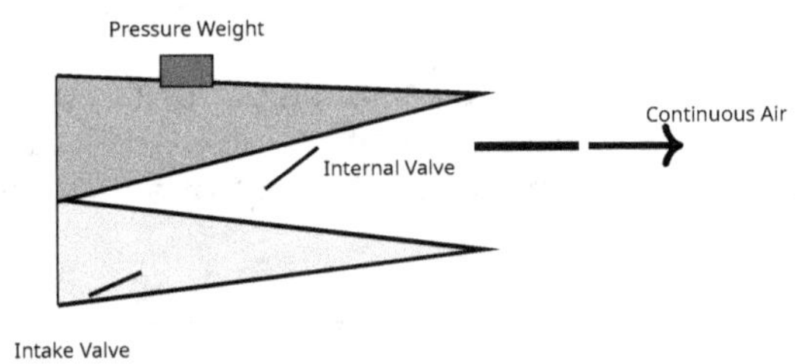

Figure 1: The internal valve and weighted reservoir ensure a non-pulsating air supply.

Pulse Elimination

Fluid dynamics in the forge environment dictate that steady oxygen delivery results in efficient carbon combustion. The double-action mechanism effectively eliminates the time intervals without airflow associated with the return stroke of a manual pump.

The air flow rate remains consistent because the energy stored in the weighted top board is released during the intake phase. This constant pressure prevents the backflow of furnace gases into the bellows, protecting the leather components from heat damage and ensuring a high-purity air blast.

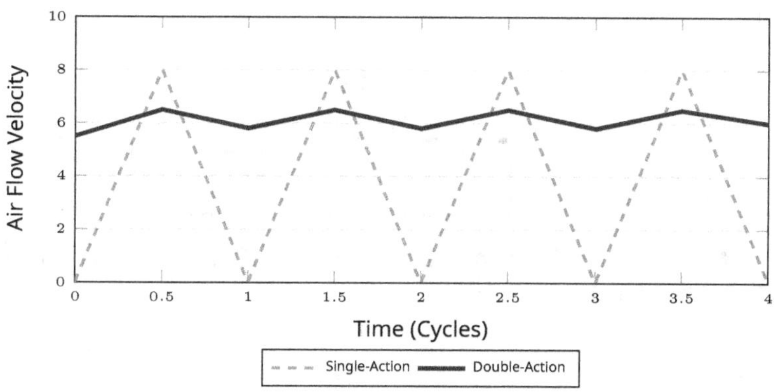

Figure 2: Output velocity comparison showing the stabilization provided by the reservoir.

Thermal Thresholds

A continuous air blast allows for the reaching of critical metallurgical temperatures. Sustained airflow enables the forge to maintain temperatures exceeding $1200°$C ($2192°$F), necessary for the welding of wrought iron and the melting of copper alloys.

The nozzle, or **tuyere**, is the point of maximum thermal stress. It is constructed of thick clay or iron to resist the intense radiant heat of the charcoal bed. The positioning of the tuyere relative to the fuel bed determines the size of the hot zone in the forge.

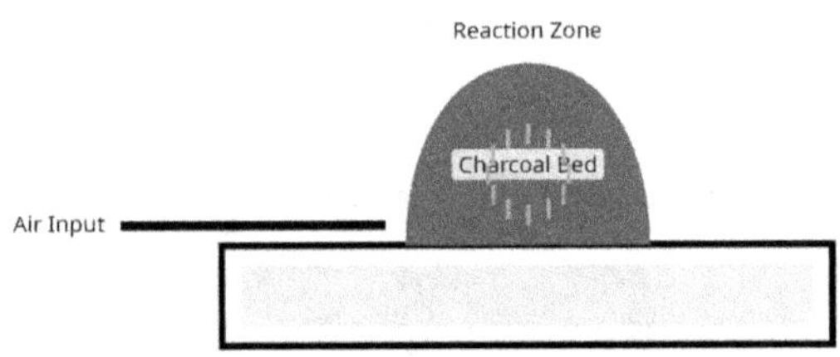

Figure 3: Concentrated airflow at the base of the fuel bed maximizes oxygen-fuel contact.

Automation of Air Supply

Industrial-scale bellows require mechanical power. Integrating the bellows with a water wheel or a heavy lever system allows for consistent operation over long periods. A camshaft mechanism converts the rotational motion of a wheel into the reciprocating motion required to drive the bellows' pump board.

The use of mechanical power increases the volume of air delivered, enabling the operation of large blast furnaces for iron production. This capability is a requirement for the transition from workshop-scale forging to industrial-scale metal production.

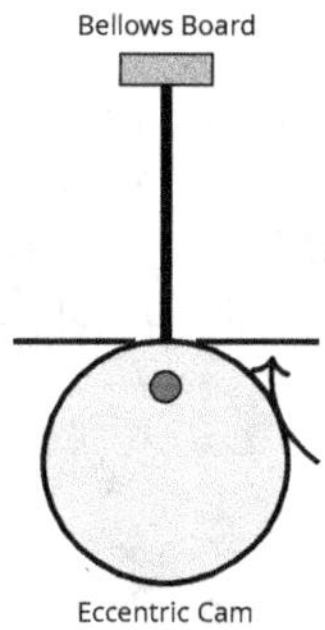

Figure 4: Camshaft integration automates the compression cycle for industrial use.

Summary

The double-action bellows represents the transition to steady-state chemical processing. By providing a continuous air supply, it enables the high-temperature environments necessary for advanced metallurgy and precision metalwork.

End of Discovery #029

DISCOVERY #030: STEAM ENGINE

Category: Physics / Power **Status:** Essential

The Vacuum Principle

The primary challenge of industrializing a civilization is the generation of mechanical power independent of animal or water resources. The steam engine utilizes the phase change of water to create a vacuum. In the Newcomen style atmospheric engine, steam is introduced into a cylinder, displacing the air.

When cold water is injected into the cylinder, the steam condenses instantly, creating a partial vacuum. The external atmospheric pressure then pushes the piston downward. This downward stroke provides the mechanical force required to operate heavy mine pumps or forge hammers.

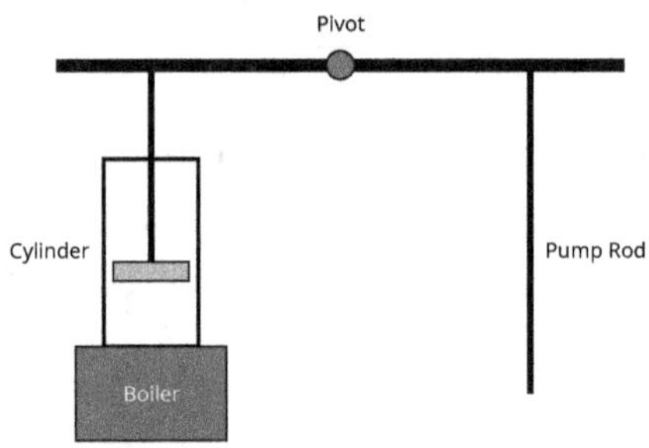

Figure 1: The beam engine uses atmospheric pressure to perform mechanical work.

Efficiency and Heat Management

The atmospheric engine consumes vast quantities of fuel because the cylinder undergoes cooling and reheating during every cycle. James Watt's improvement involves the addition of a **Separate Condenser**.

In this configuration, the main cylinder remains constantly hot, while the condensation occurs in a connected, smaller vessel that remains constantly cold. This prevents the loss of thermal energy required to reheat the large mass of the main piston and cylinder walls, increasing the engine's thermal efficiency.

Valve Coordination

Automated valves coordinate the flow of steam. The intake valve allows steam into the hot cylinder, while the exhaust valve leads to the cold condenser, maintaining the pressure differential necessary for continuous motion.

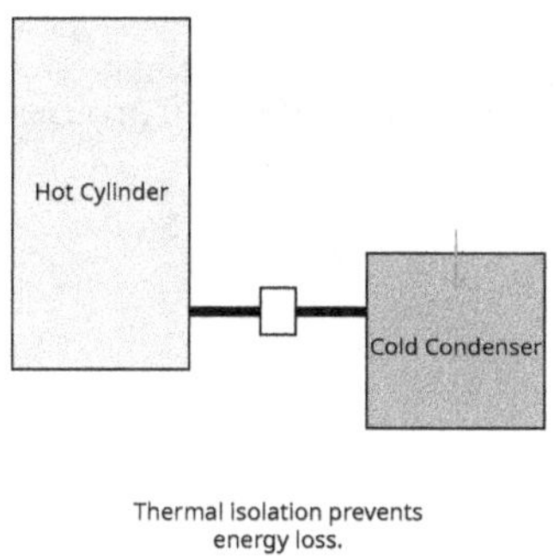

Figure 2: The separate condenser maintains a constant thermal gradient.

The Work Cycle

The efficiency of a steam engine is visualized using the **Pressure-Volume (P-V) Diagram**. This graph plots the internal pressure of the cylinder against the volume occupied by the steam throughout a single piston stroke.

The area enclosed by the curve represents the net work performed by the engine in one cycle. Optimization involves the early cutoff of steam, allowing the steam to expand and perform work after the intake valve is closed, utilizing the internal energy of the vapor more effectively.

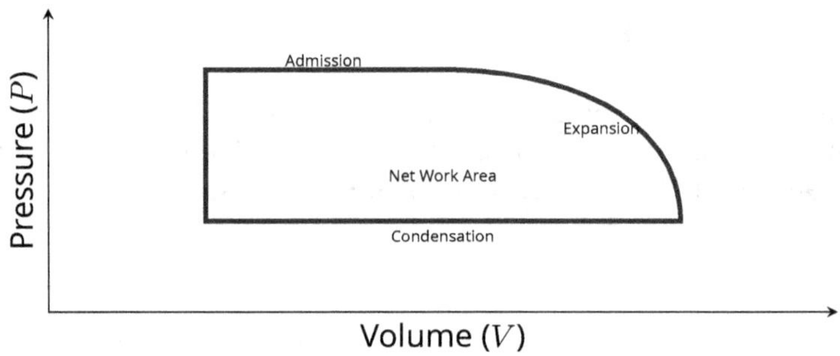

Figure 3: The enclosed area of the P-V diagram defines the mechanical output.

Conversion of Power

To drive machinery such as lathes, looms, and millstones, reciprocating motion must be converted into **rotary motion**. This is achieved through the integration of a connecting rod, a crank, and a heavy flywheel.

The **flywheel** acts as a kinetic energy reservoir. Due to its high rotational inertia, it smooths out the power pulses of the engine, ensuring a constant rotational speed despite variable loads. This stability is essential for precision manufacturing and the safe operation of high-speed industrial tools.

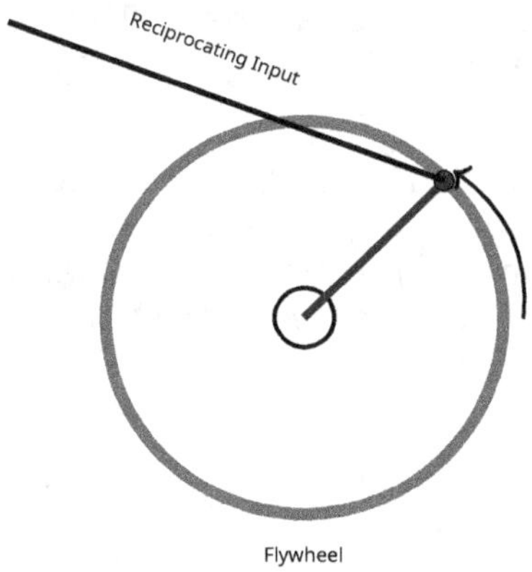

Figure 4: Rotational inertia enables smooth power delivery for complex machines.

Summary

The steam engine initiates the thermal-to-mechanical energy conversion era. By decoupling power generation from natural sites, it allows for the centralization of industry and the development of self-powered transportation systems.

End of Discovery #030

DISCOVERY #031: THE WATER TURBINE

Category: Fluid Mechanics / Power **Status:** Essential

The Kinetic Energy Transfer

The water turbine is a rotary machine that extracts energy from a moving fluid. In an impulse turbine, a high-velocity jet of water strikes curved buckets mounted on the periphery of a wheel. The change in the fluid's momentum creates a force on the buckets, resulting in high-torque rotational motion.

Efficient energy transfer occurs when the bucket velocity is approximately half the velocity of the incoming water jet. This relationship ensures that the water leaves the bucket with minimal residual velocity, transferring the maximum amount of kinetic energy to the shaft.

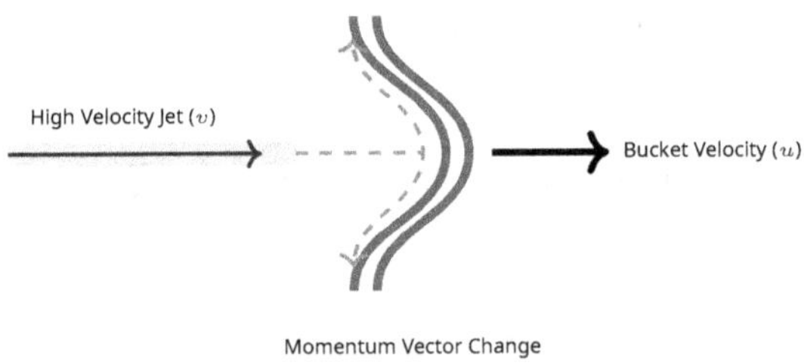

Figure 1: Directional reversal of the fluid jet maximizes mechanical work.

Bifurcated Geometry

The Pelton-style bucket utilizes a central splitter edge to divide the water jet into two equal streams. This design balances the lateral forces on the wheel assembly, prevents axial thrust, and reduces wear on the bearings.

The internal curvature of the buckets turns the water nearly 180 degrees. This specific geometry facilitates high efficiency by utilizing the full reactionary force of the water as it is decelerated and redirected.

Nozzle Regulation

The volume of water delivered is regulated by a needle nozzle. Adjusting the position of the spear within the nozzle allows for precise control of the jet diameter while maintaining high jet quality at variable loads.

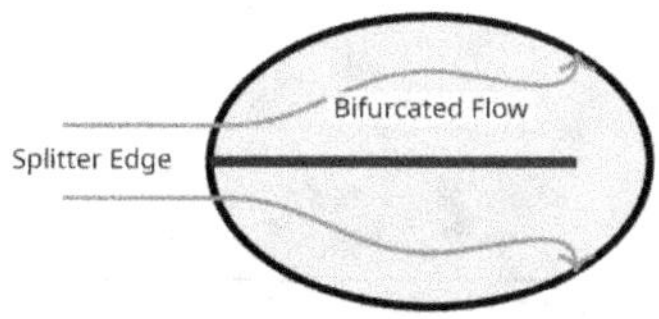

Figure 2: The splitter divides the impulse force evenly across the turbine wheel.

Speed Ratio Optimization

Turbine performance is governed by the ratio of bucket speed to jet speed (u/v). A stationary wheel ($u = 0$) produces maximum torque. Work output is zero in this state as there is no displacement. A wheel spinning at the speed of the jet ($u = v$) produces no torque.

Theoretical peak efficiency occurs at a speed ratio of 0.5. In industrial practice, Pelton turbines achieve efficiencies exceeding 90% across operating conditions. This performance makes the turbine the ideal engine for high-head water sources, where water pressure is converted into high-velocity kinetic energy.

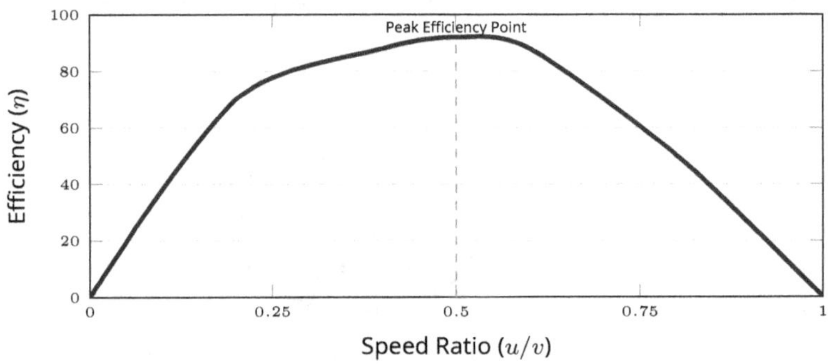

Figure 3: Output work peaks when the bucket speed is half the water velocity.

Mechanical Integration

The high rotational speed of the turbine shaft allows for direct coupling to industrial machinery. Turbines operate with high precision and low mechanical vibration.

The turbine shaft provides the power necessary for:

- **High-Speed Milling:** Efficient flour production and textile processing.
- **Compressed Air Systems:** Powering pneumatic tools in workshops.
- **Electrical Generation:** Driving dynamos for localized lighting and chemistry.

Turbines enable the exploitation of high-pressure mountain streams, providing a concentrated power source. This technology provides energy densities higher than those of traditional gravity-fed wheels.

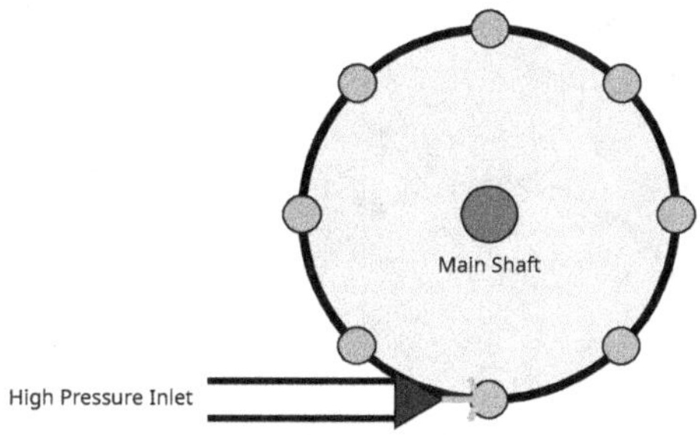

Figure 4: The compact assembly enables efficient power extraction from fluid volumes.

DISCOVERY #032: SPINNING WHEEL

Category: Textiles / Mechanics **Status:** Essential

The Principle of Torque Conversion

Spinning is the process of twisting fibers together to form high-strength yarn. The spinning wheel utilizes mechanical advantage to increase the rotational velocity of the spindle. A large drive wheel connects to a small pulley, or whorl, by a tensioned band.

A single rotation of the drive wheel results in multiple rotations of the spindle. This high-speed rotation provides the torque necessary to align and twist raw fibers—such as wool, flax, or cotton—into standardized threads. The spinning wheel achieves a production rate higher than manual drop-spinning methods.

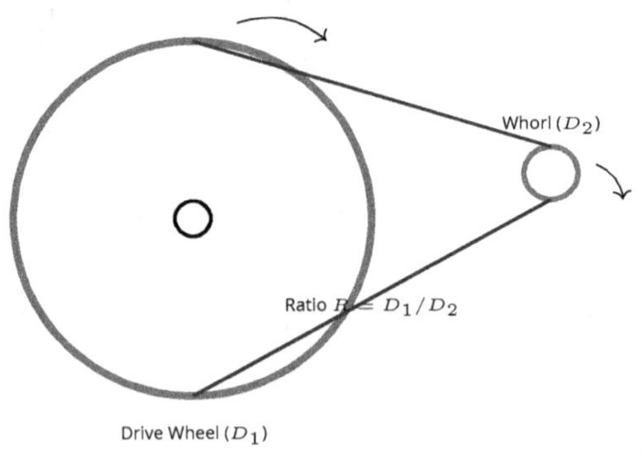

Figure 1: The pulley system multiplies rotational speed for efficient spinning.

Simultaneous Twist and Wind

Advanced spinning wheels utilize a flyer mechanism to perform fiber twisting and winding of yarn onto a bobbin simultaneously. The flyer is a U-shaped component that rotates around the bobbin.

Differential speed between the flyer and bobbin is managed by a friction brake or a separate drive band. As the flyer twists the yarn, the lag in rotation relative to the bobbin causes the yarn to wind onto the storage cylinder. This continuous operation removes the requirement to stop the machine during winding.

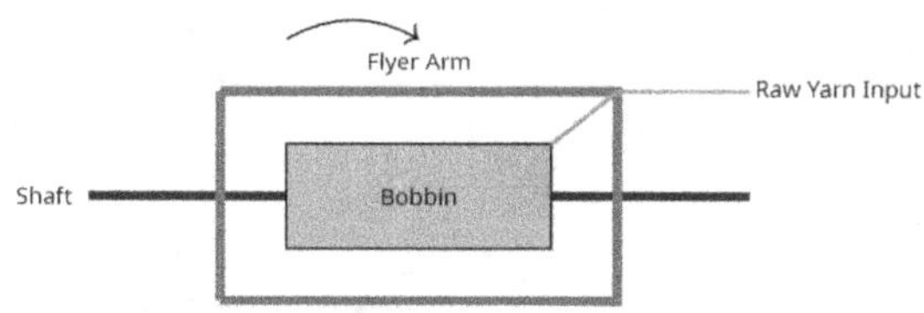

Figure 2: The flyer mechanism integrates fiber twisting and automated winding.

Output Standardization

Mechanical spinning produces yarn of uniform thickness and twist density. Uniformity is essential for high-tension applications, such as the production of heavy canvas for sails or industrial clothing.

The productivity increase provided by the wheel allows for the processing of large volumes of raw material. This efficiency enables a single operator to supply multiple weavers, facilitating the transition to workshop-scale textile manufacturing.

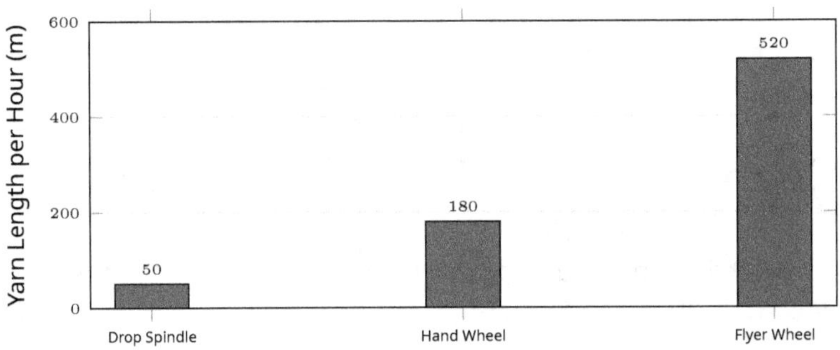

Figure 3: Flyer-equipped wheels provide a tenfold increase in yarn production capacity.

Industrial Scaling

Thread standardization allows for the development of complex looms. Mechanical weaving requires consistent yarn strength. Manual weaving methods permit the correction of irregularities during the process, whereas mechanical systems depend on uniform stock.

Strategic Impact:

- **Maritime Expansion:** Rapid production of high-durability sailcloth.
- **Standardized Uniforms:** Protective garments for industrial labor.
- **Trade Goods:** Surplus textiles serve as a currency for inter-regional commerce.

The spinning wheel transforms a manual craft into a high-throughput industrial process, forming the basis of the textile economy.

Fiber Type	Typical Ratio	RPM at 60 Wheel RPM
Coarse Wool	10:1	600
Medium Cotton	15:1	900
Fine Flax/Linen	25:1	1,500

Table 1: Spindle rotation requirements for different industrial fibers.

Summary

The spinning wheel introduces precise mechanics to textile production. By multiplying operator input through pulley systems and automated winding, a civilization ensures the mass production of fabrics required for industry and survival.

End of Discovery #032

DISCOVERY #033: THE MECHANICAL LOOM

Category: Textiles / Mechanics **Status:** Essential

The Geometry of Fabric

Weaving is the process of interlacing two distinct sets of threads at right angles to create a stable fabric. The longitudinal threads, known as the **warp**, are held in tension across the frame of the machine. The transverse thread, known as the **weft**, is passed through the warp to form the structure of the textile.

Fabric stability depends on the pattern of interlacing. The plain weave provides the highest number of interlacing points per unit area. In this configuration, the weft passes over and under alternating warp threads, resulting in a durable and balanced cloth.

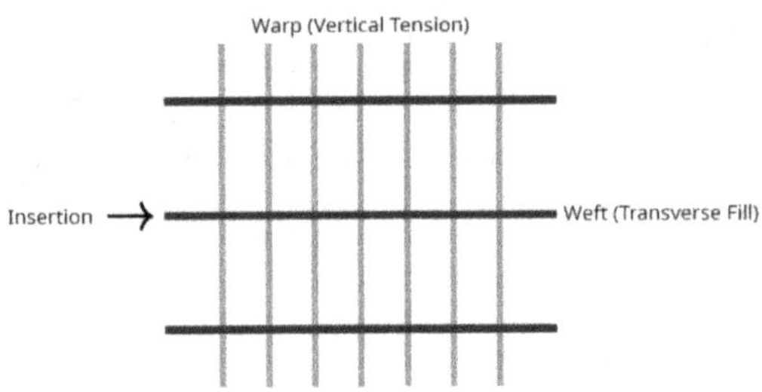

Figure 1: The interlacing of perpendicular fibers forms the structural lattice of textiles.

The Shedding Cycle

The mechanical loom automates the separation of warp threads through a process called **shedding**. Warp threads are threaded through individual eyelets in vertical frames known as heddles. Raising specific heddles creates a triangular opening, or the shed.

This opening allows the weft to pass through the warp in a single motion. In a mechanical system, the movement of heddles is controlled by a camshaft. This synchronization ensures that thread separation occurs simultaneously with the insertion of the weft, removing the requirement for manual manipulation.

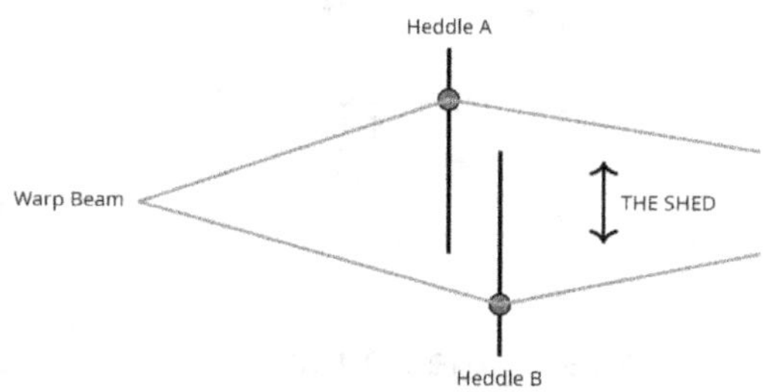

Figure 2: The rhythmic lifting of heddles creates the path for weft insertion.

Picking and Battening

Weft thread insertion, or **picking**, uses a shuttle. The flying shuttle is an aerodynamic housing for the weft bobbin equipped with wheels. A striking mechanism propels it through the shed at high velocity.

Following each pick, the **reed** (a comb-like structure) moves forward to push the new weft thread against the woven cloth. This process, known as **battening**, ensures a high-density, uniform weave. Integrating shedding, picking, and battening allows for continuous, high-speed textile production.

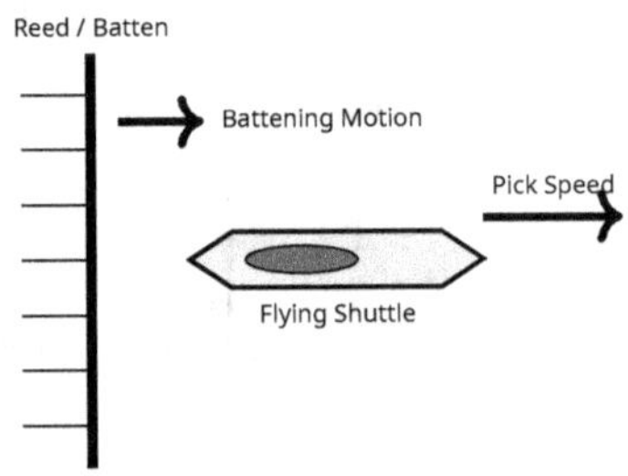

Figure 3: The reed compresses the weft threads to maintain fabric density.

Industrial Scaling

The mechanical loom increases weaver productivity by an order of magnitude. The physical reach of the operator and the speed of hand-passing the shuttle constrain manual weaving. Mechanical drives enable wider fabrics and a significantly higher frequency of picks per minute.

Constant tension systems and automated take-up rollers wind the finished cloth onto a beam at a uniform rate. Mechanical consistency allows the production of heavy-duty textiles required for industrial belts, sailcloth, and standardized protective clothing.

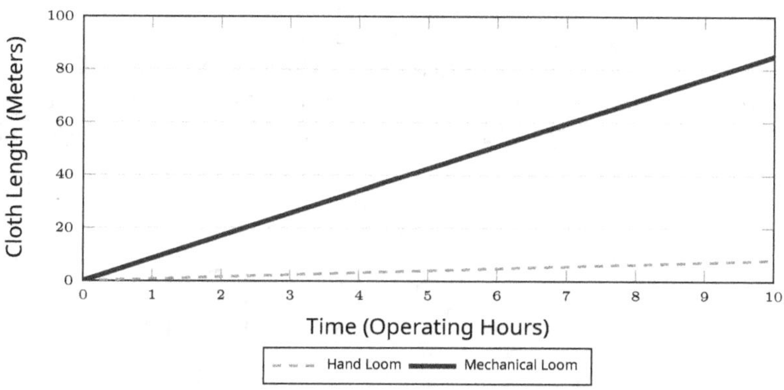

Figure 4: Mechanical automation decouples production speed from human physical limits.

Summary

The mechanical loom transforms textile production from a skilled manual craft into a high-capacity industrial process. By synchronizing the fundamental motions of weaving, a civilization gains the ability to mass-produce the materials essential for housing, clothing, and infrastructure.

End of Discovery #033

DISCOVERY #034: THE ROCKET STOVE

Category: Thermal Engineering / Energy **Status:** Essential

The Efficient Burn

The rocket stove is a high-performance heating system designed to achieve near-complete combustion of wood fuel. It operates on two primary principles: high-temperature insulated burning and a strong constant draft.

Efficient combustion requires an insulated burn tunnel to keep the combustion zone hot. This temperature is sufficient to burn volatile gases and carbon particles that escape as smoke in traditional fires. The resulting clean combustion extracts maximum thermal energy from minimal wood volumes.

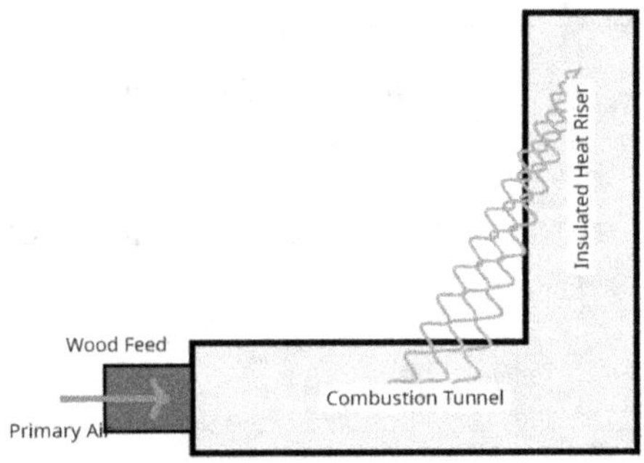

Figure 1: The L-shaped insulated chamber facilitates high-temperature combustion.

Thermosyphon and Draft

The vertical heat riser is the primary driver of the system. High levels of insulation maintain a temperature differential between internal hot gases and surrounding air. This differential creates a powerful upward draft via the chimney effect.

This strong draft pulls primary air into the firebox, providing a continuous oxygen supply. The system generates this draft internally and functions without external power or fans. Exhaust gas velocity ensures the fire remains concentrated at the base of the riser.

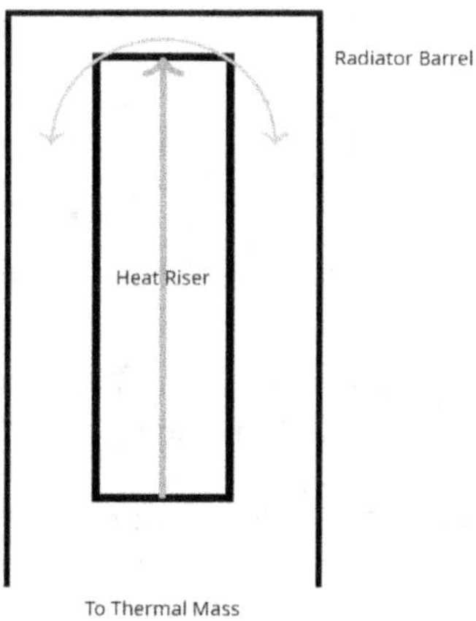

Figure 2: The riser forces hot gases to circulate against the outer radiator surface.

Heat Exchange and Retention

Efficient heating requires thermal energy extraction from exhaust before it exits the structure. Rocket Mass Heaters duct hot gases through a large **thermal mass** constructed of cob or heavy masonry.

The thermal mass absorbs heat from moving gases through conduction and radiates stored energy slowly over 12 to 24 hours. This process decouples combustion from heating, allowing a short, intense fire to heat a living space for an entire day.

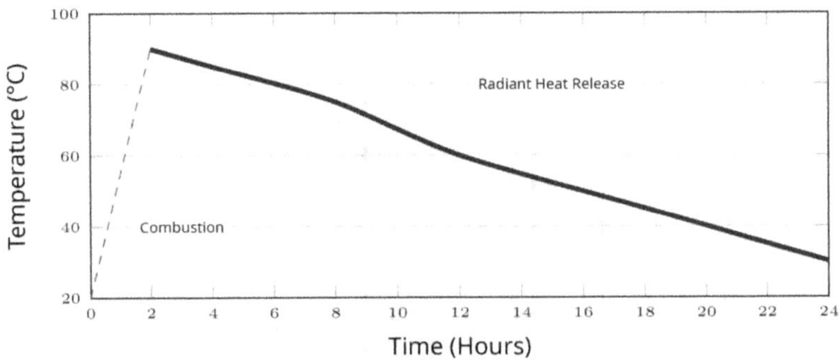

Figure 3: Thermal mass acts as a chemical energy capacitor for domestic heating.

Construction and Materials

The rocket stove is constructed from locally available refractory materials. The core withstands temperatures exceeding $1000°$C without structural failure.

- **The Core:** Firebricks or high-clay cob mixes provide heat resistance.
- **Insulation:** Perlite or wood ash packed around the riser prevents heat loss to the mass.
- **The Manifold:** A heavy-gauge metal barrel or stone chamber acts as the primary radiator.
- **The Bench:** The mass is often shaped into a seating area for direct conductive heating.

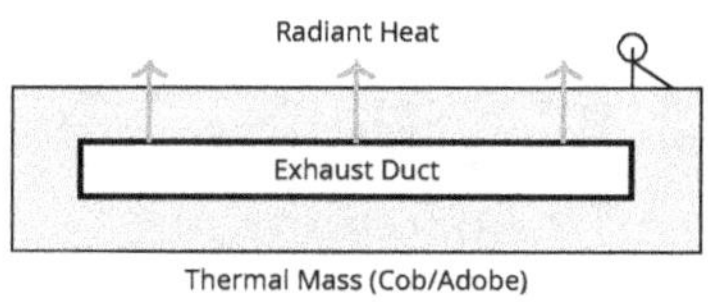

Figure 4: The horizontal ducting maximizes contact time between hot gases and storage material.

Summary

The rocket stove represents a fundamental advancement in thermal management. By ensuring complete combustion and maximizing heat retention, a civilization maintains stable heating environments with a fraction of the fuel required by traditional methods.

End of Discovery #034

DISCOVERY #035: ILLUMINATION

Category: Energy / Optics **Status:** Essential

Combustion Dynamics

Illumination results from controlled chemical combustion. A candle or oil lamp functions as a self-regulating gas generator. Heat from the flame liquefies the fuel—tallow, wax, or oil—which is then drawn upward into the wick via capillary action.

Within the wick, the liquid fuel undergoes pyrolysis, breaking down into volatile gases. These gases react with atmospheric oxygen in the combustion zone. The visible light is produced by incandescent carbon particles (soot) that glow due to the high temperature of the reaction.

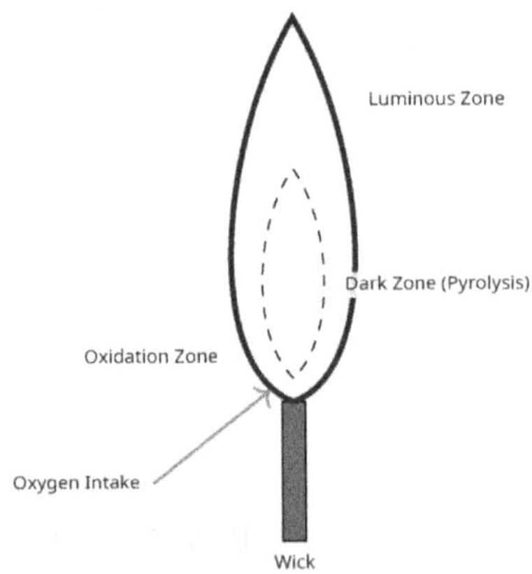

Figure 1: The chemical zones of a diffusion flame enable light production.

Tallow and Lipids

Fuel selection determines the duration and cleanliness of the light source. Tallow, derived from rendered animal fat, consists of triglycerides. It provides a reliable fuel source. Refining tallow involves boiling it in water to remove cellular debris.

Adding a small percentage of beeswax increases the melting point, resulting in a firmer structure that resists guttering. Waxes are esters of long-chain fatty acids. Their high carbon-to-hydrogen ratio ensures high luminosity. Illumination requires thermal energy to initiate the release of stored chemical energy.

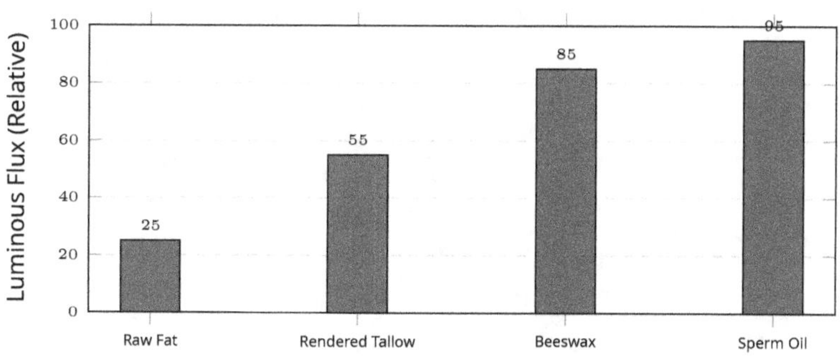

Figure 2: Fuel refinement increases the conversion efficiency of chemical energy to light.

Oxygen Regulation

Efficient combustion requires precise oxygen delivery. The Argand lamp utilizes a circular, hollow wick and a glass chimney to manage airflow. This configuration allows air to flow through the center of the wick and along its outer surface simultaneously.

The glass chimney creates a steady upward draft, increasing the velocity of the air intake. This increased oxygen supply raises the combustion temperature, resulting in a brighter and cleaner light.

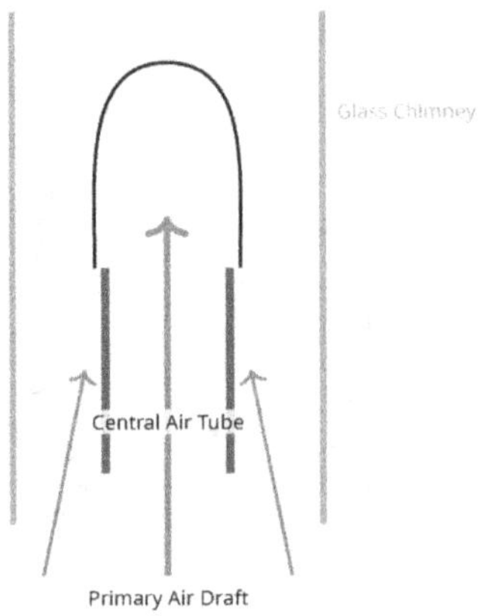

Figure 3: The central draft increases oxygen availability, eliminating smoke.

Reflection and Concentration

Effective illumination requires directing light toward the area of utility. According to the inverse square law, light intensity decreases rapidly with distance.

Concentrating light into a parallel beam is achieved through a parabolic reflector. The light source is positioned at the geometric focus of the parabola. Rays striking the mirror surface reflect along a single axis, maximizing the effective range of the source. This optical configuration is essential for lighthouses and signaling systems.

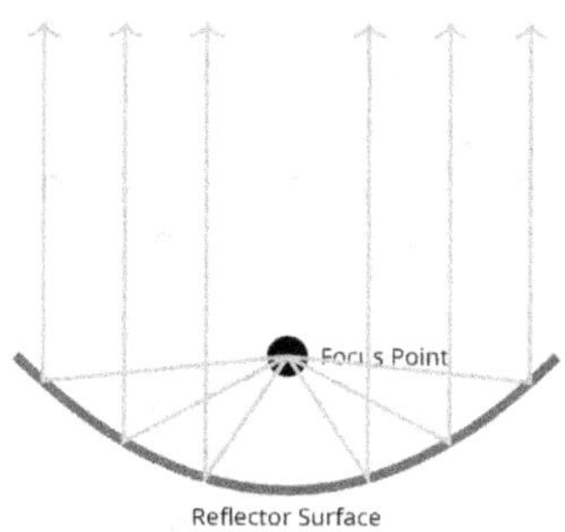

Figure 4: Parabolic geometry redirects divergent rays into a concentrated beam.

Summary

Illumination technology shifts from simple combustion to engineered thermal systems. By managing oxygen flow and utilizing optical reflection, a civilization gains control over its environment, enabling continuous activity.

End of Discovery #035

DISCOVERY #036: THE MICROSCOPE

Category: Optics / Biology **Status:** Essential

The Refraction of Light

The microscope utilizes the refractive property of glass to bend light rays. When light passes from air into a convex lens, its velocity changes, causing the rays to converge toward a focal point.

Lenses are characterized by their focal length. A lens with a short focal length provides high magnifying power. Positioning an object just beyond the focal point of a lens produces a magnified, inverted image. This discovery allows for the visualization of structures far below the resolution limit of the human eye.

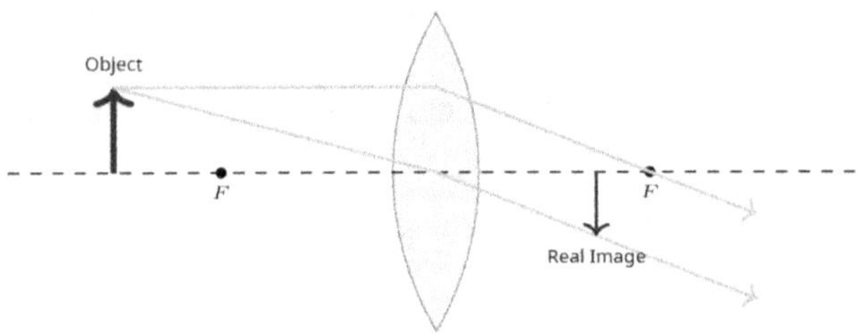

Figure 1: Convex lenses converge light rays to create magnified representations.

Secondary Magnification

The compound microscope employs a two-stage optical system. The primary lens, the objective, creates a magnified image of the specimen. The secondary lens, the eyepiece, magnifies this primary image a second time.

The total magnification of the system is the product of the powers of both lenses. This dual-lens arrangement enables significantly higher magnification levels than a single-lens glass. Precise focus is maintained through a screw-threaded mechanism that adjusts the distance between the lenses and the specimen.

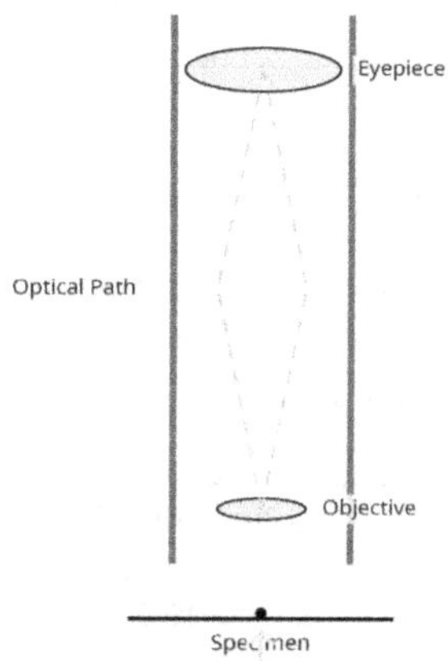

Figure 2: Optical components are aligned within a tube to isolate the light path.

Resolution and Wave Physics

Magnification increases the size of the image, while resolution defines the ability to distinguish between two closely spaced points. Resolution is governed by the wavelength of light and the numerical aperture of the objective lens.

Increased magnification without a corresponding increase in resolution results in empty magnification—a larger image that provides no additional detail. Optimizing illumination through a condenser lens allows for the visualization of objects as small as 0.2 micrometers. This technical limit is fundamental to biological observation.

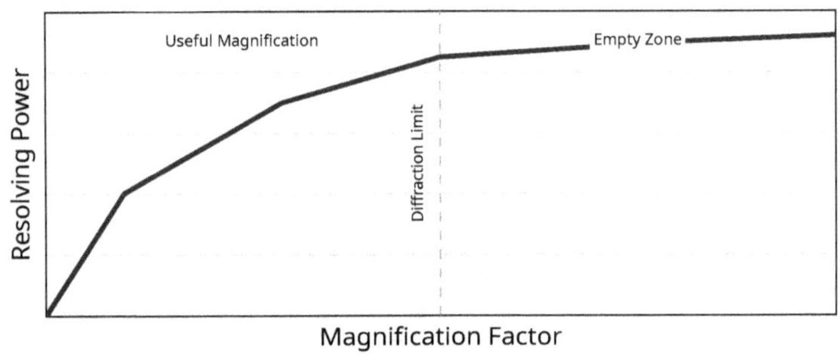

Figure 3: Useful magnification is capped by the wave properties of light.

Germ Theory and Microbiology

The microscope provides direct evidence of microorganisms. Many diseases result from microscopic pathogens—bacteria, fungi, and protozoa. Observing the morphology of these organisms allows for their identification and the implementation of sterilization protocols. This identification is required for effective surgery and public sanitation.

Primary Microbial Forms:

- **Cocci:** Spherical bacteria (e.g., Staphylococci).
- **Bacilli:** Rod-shaped bacteria (e.g., Anthrax).
- **Spirilla:** Spiral-shaped bacteria (e.g., Cholera).

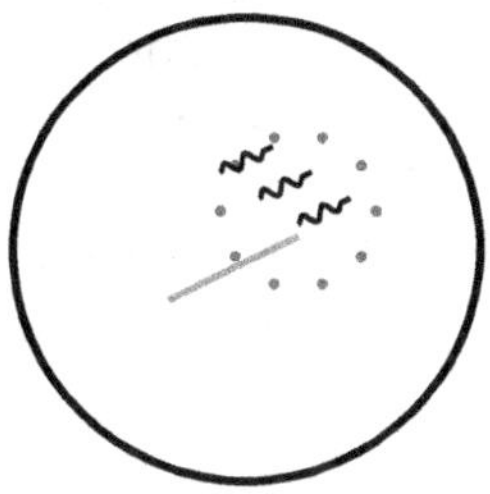

Microbial Morphology

Figure 4: Direct observation confirms the presence of pathogenic agents.

Summary

The microscope transforms medicine into an exact science. By exposing the hidden biological world, it enables the prevention of infections and the systematic study of cellular life, establishing the foundation for modern health.

End of Discovery #036

DISCOVERY #037: ANTISEPSIS

Category: Medicine / Biology **Status:** Essential

The Principle of Asepsis

Antisepsis is the destruction of pathogenic microorganisms on living tissue or medical instruments. Successful medical intervention requires the maintenance of a sterile field to prevent post-operative infections.

Inactivation occurs through the mechanical and chemical disruption of microbial structures. High temperatures and specific chemical agents target proteins and lipid membranes. These agents render bacteria and viruses unable to replicate. The effectiveness of a sterilization process is defined by the logarithmic reduction of the microbial population over a specific duration of exposure.

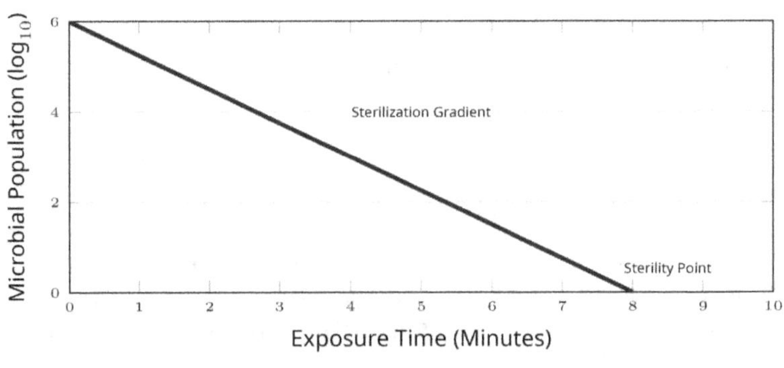

Figure 1: The relationship between exposure duration and pathogen survival.

Alcohol and Denaturation

Ethanol (C_2H_5OH) and Isopropyl alcohol are primary antiseptic agents. An aqueous solution of 70% alcohol is the standard for skin preparation.

The mechanism of action involves the denaturation of proteins. Alcohol molecules penetrate the cell wall and unfold protein chains. This causes coagulation and a loss of biological activity. Water acts as a catalyst in this process. Pure 100% alcohol causes rapid surface dehydration, creating a protective shell that prevents the agent from reaching internal microbial structures.

Phenolic Compounds

Carbolic acid (phenol) serves as a disinfectant for surgical environments. Spraying a dilute solution of phenol in operating rooms reduces airborne pathogen concentrations, lowering infection rates in open wounds.

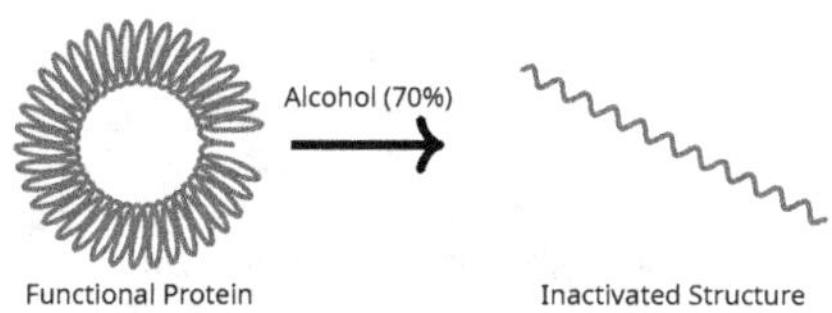

Figure 2: Chemical agents structurally dismantle the biological machinery of pathogens.

Moist Heat and Pressure

Heat is a reliable method for sterilizing metal tools and textiles. Boiling water at $100°$C kills vegetative bacteria. Highly resistant endospores require higher temperatures for total destruction.

The **Autoclave** utilizes pressurized steam to reach temperatures exceeding the atmospheric boiling point of water. At a gauge pressure of 1 bar (15 psi), steam reaches approximately $121°$C ($250°$F). Exposure to this environment for 20 minutes ensures the destruction of all known biological agents through protein coagulation and DNA damage.

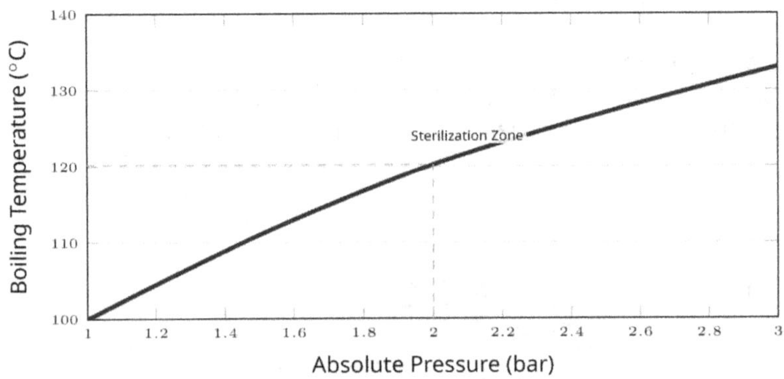

Figure 3: Increased pressure raises the steam temperature to biocidal levels.

Debridement and Cleaning

Mechanical cleaning is the prerequisite for antiseptic application. Dirt, necrotic tissue, and foreign objects protect pathogens from chemical agents.

Irrigation: Flushing the wound with sterile saline solution removes contaminants. **Debridement:** Surgical removal of dead tissue prevents the spread of bacteria into healthy margins.

Clean water, sterile dressings, and chemical cleansing form the basis of wound management. These steps prevent gangrene and localized infections.

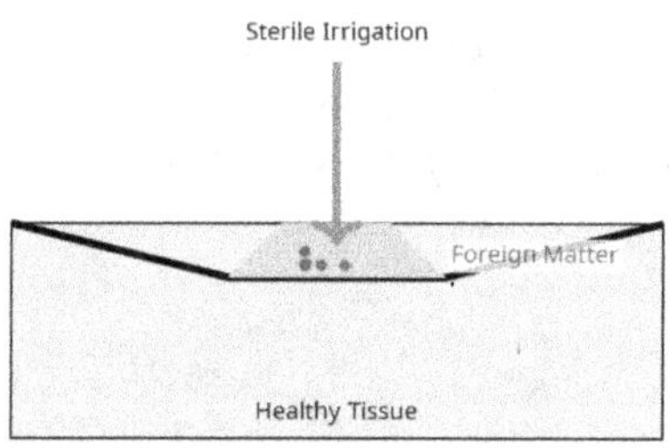

Figure 4: The mechanical removal of debris is essential for successful antisepsis.

Summary

Antisepsis and sterilization establish a controlled biological environment. By utilizing chemical denaturation and pressurized thermal energy, a civilization eliminates the primary cause of post-operative mortality.

End of Discovery #037

DISCOVERY #038: ANESTHESIA

Category: Medicine / Chemistry **Status:** Essential

Suppression of Pain Signals

Anesthesia is the pharmacological induction of a reversible state of unconsciousness. In a surgical context, it prevents physiological shock and trauma.

The mechanism involves interference with neural transmission. Anesthetic molecules interact with the lipid bilayer of nerve cell membranes or specific protein receptors. This interaction inhibits the movement of ions across the membrane. This inhibition prevents the generation of electrical signals that the brain interprets as pain.

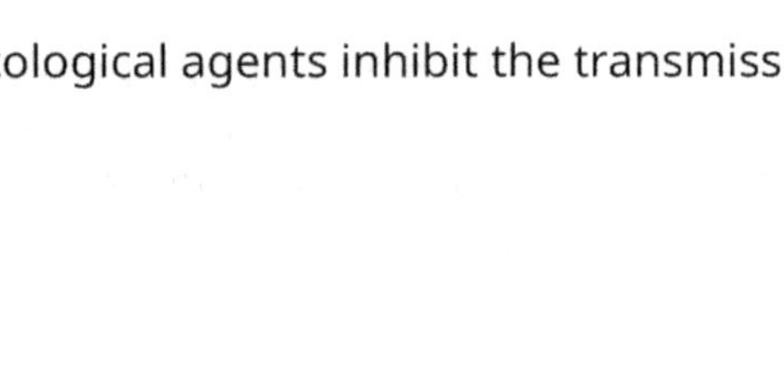

Figure 1: Pharmacological agents inhibit the transmission of pain stimuli.

Ether Synthesis

Diethyl ether ($C_4H_{10}O$) is a volatile liquid produced through the dehydration of ethanol. Concentrated sulfuric acid acts as the catalyst for this reaction.

The process requires heating the mixture to approximately $140°$C ($284°$F). Sulfuric acid removes a water molecule from two ethanol molecules to create an ether linkage. The resulting vapor is condensed and collected. Administered via inhalation, ether induces deep sleep and muscle relaxation while maintaining a safety margin for respiratory function.

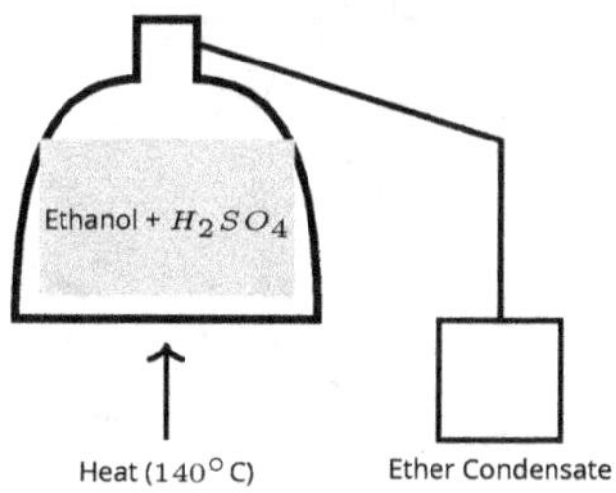

Figure 2: The catalytic dehydration of alcohol yields volatile diethyl ether.

The Haloform Reaction

Chloroform ($CHCl_3$) is a heavy, colorless liquid produced by reacting ethanol or acetone with chlorinated lime.

Chloroform is a potent anesthetic and requires a low concentration to induce unconsciousness. It is non-flammable. This property makes it safe for use near open flames. Administration requires precise dosage control. The therapeutic window is narrow, and excessive concentrations cause cardiac arrhythmias.

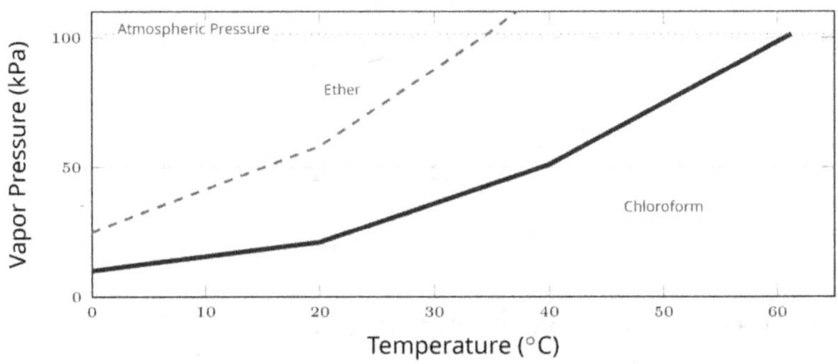

Figure 3: The high vapor pressure of haloforms enables inhalation delivery.

The Alkaloids of Opium

Opium, derived from the *Papaver somniferum* poppy, contains natural alkaloids such as morphine. These compounds act on the central nervous system to provide profound analgesia.

Morphine binds to opioid receptors in the brain and spinal cord. This binding inhibits the release of neurotransmitters involved in pain signal transmission. Inhalation anesthetics induce unconsciousness. Opiates provide systemic analgesia during and after surgery. Their dosage requires monitoring to prevent respiratory depression.

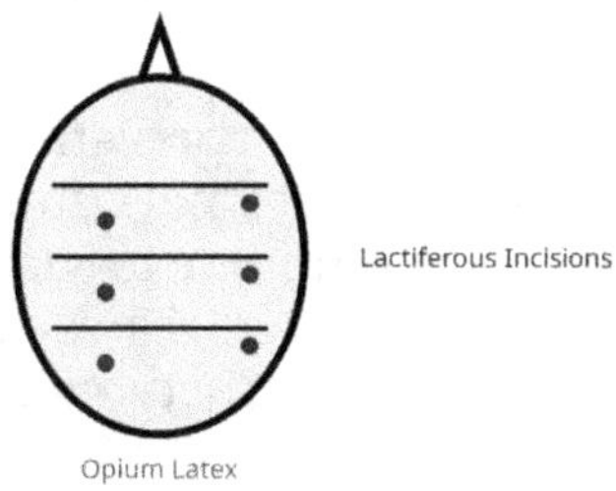

Figure 4: Natural alkaloids provide the chemical basis for potent systemic analgesia.

Summary

Anesthesia removes the primary obstacle to surgery. By suppressing the nervous system response to trauma, a civilization enables procedures that would be impossible due to pain-induced shock.

End of Discovery #038

DISCOVERY #039: NATURAL ANTIBIOTICS

Category: Medicine / Biochemistry **Status:** Essential

The Mechanism of Action

Antibiotics are secondary metabolites produced by microorganisms to inhibit the growth of competing bacteria. Penicillin, derived from *Penicillium* molds, interferes with the synthesis of **peptidoglycan**, a critical component of the bacterial cell wall.

During bacterial division, the antibiotic prevents the formation of cross-links in the cell wall structure. The internal osmotic pressure of the bacterium causes the weakened cell membrane to rupture. This process, known as **bacteriolysis**, results in cell death. Penicillin action is specific to bacterial cells. Human cells lack peptidoglycan structures and remain unaffected by the compound.

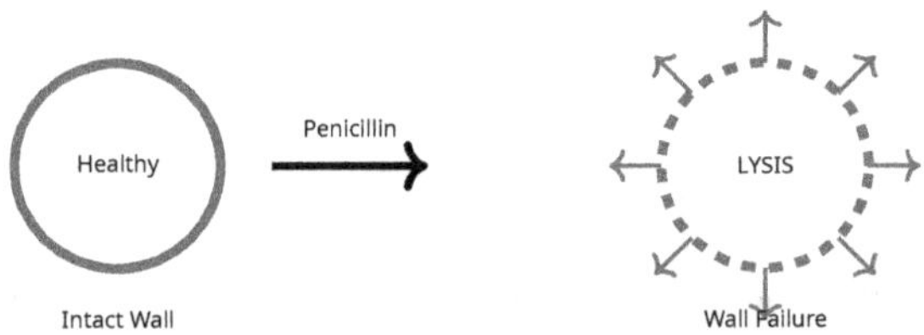

Figure 1: Molecular interference leads to terminal osmotic rupture of the cell.

Penicillium Morphology

Identification of the correct fungal strain is the requirement for antibiotic production. *Penicillium chrysogenum* produces blue-green brush-like spore structures known as conidiophores. These molds grow on organic substrates including bread, citrus fruit, and specialized nutrient broths.

Production involves surface culture or deep-tank fermentation in sterilized vessels. The mold utilizes carbon sources such as glucose and nitrogen sources like corn steep liquor to synthesize the active penicillin molecule. Maintaining precise temperature and aeration levels maximizes the yield of the compound.

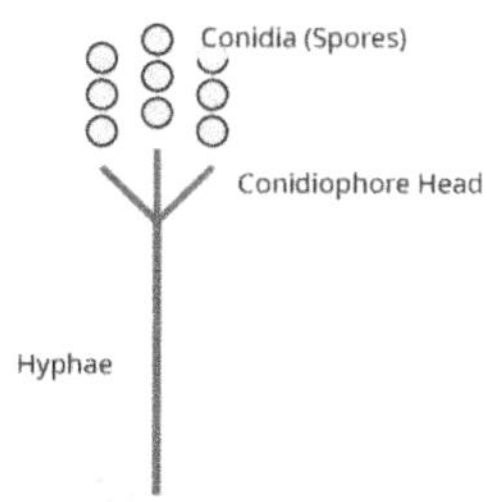

Figure 2: The distinctive morphology of *Penicillium* conidiophores identifies the strain.

Osmotic Pressure Mechanics

Honey acts as a natural topical biocide. Its primary action is high **osmotic pressure**. As a supersaturated sugar solution, honey draws water out of bacterial cells through the semi-permeable cell membrane. This dehydration effectively kills the microorganisms.

Hydrogen Peroxide Release

Honey contains the enzyme glucose oxidase. Contact with wound exudates dilutes the honey and activates the enzyme. This reaction produces a slow, sustained release of **hydrogen peroxide** (H_2O_2). This compound destroys bacterial components through oxidation without causing damage to the healthy tissue in the wound bed.

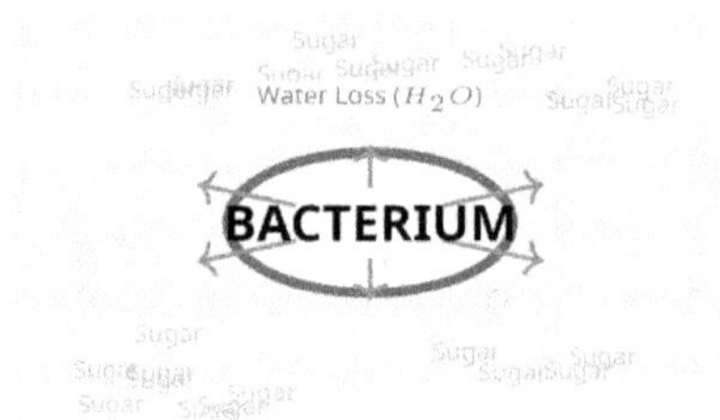

Figure 3: High solute concentrations in honey induce lethal bacterial dehydration.

Secondary Metabolite Production

Antibiotic production in a batch culture follows a consistent growth curve. Synthesis of the compound occurs during the **stationary phase**, after the period of rapid logarithmic growth. The mold produces the antibiotic as a defense mechanism when nutrient resources in the environment become limited.

Recovery involves the filtration of the mycelium from the liquid broth. The penicillin is extracted from the remaining liquid using solvents or adsorption onto activated charcoal. The final product is purified through a crystallization process to ensure medical-grade potency.

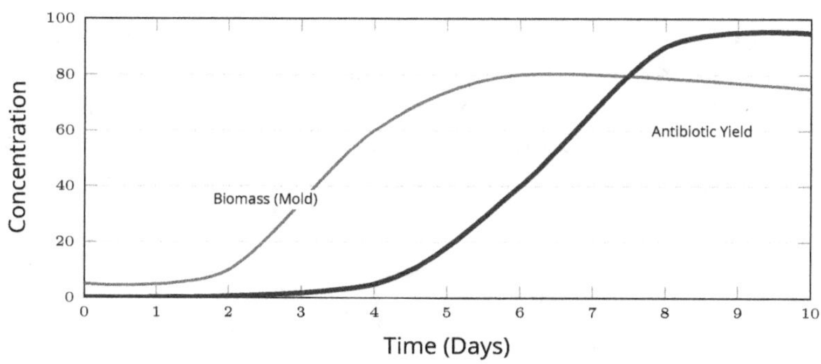

Figure 4: Antibiotic synthesis peaks during the stationary phase of mold growth.

Summary

Natural antibiotics utilize the biological defense mechanisms of microorganisms. Concentration of these compounds allows for the treatment of systemic infections, establishing the foundation of internal medicine.

End of Discovery #039

DISCOVERY #040: EMERGENCY DENTISTRY

Category: Medicine / Surgery **Status:** Essential

Mineralized Organ Structure

A tooth is a multi-layered organ adapted for mechanical processing. The **enamel** is the outermost layer and the hardest substance in the human body. The **dentin** lies beneath the enamel, providing a mineralized resilient structure.

The central **pulp** cavity contains the nerves and blood supply. Bacterial decay (caries) that breaches the enamel and dentin leads to pulp infection. This infection causes inflammation, high internal pressure, and pain. Untreated dental infections progress to abscesses and systemic sepsis.

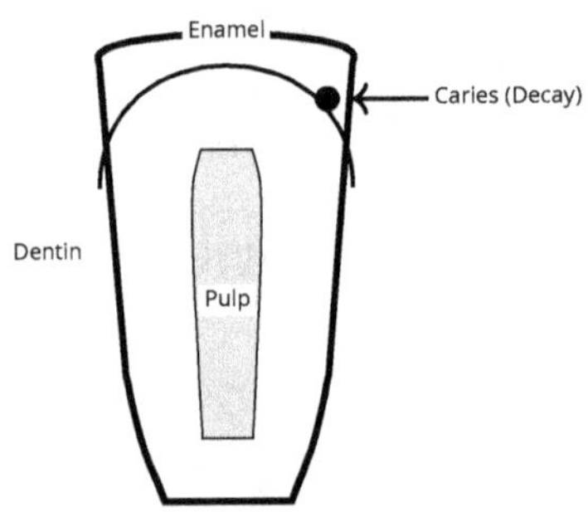

Figure 1: The breach of mineralized layers leads to internal neural infection.

Leverage and Fulcrum

Tooth extraction is necessary when repair is impossible. This surgical procedure utilizes mechanical advantage through the application of elevators and forceps.

The elevator functions as a first-class lever. Positioning the tip between the tooth root and the alveolar bone and applying a force to the handle generates the upward force required to loosen (luxate) the tooth. Forceps apply rotational and lateral forces to break the periodontal ligaments.

Surgical Safety Protocols

Controlled application of force prevents jawbone or root fractures. Sterilization of instruments using heat or alcohol is the requirement to prevent the introduction of pathogens into the bloodstream during the procedure.

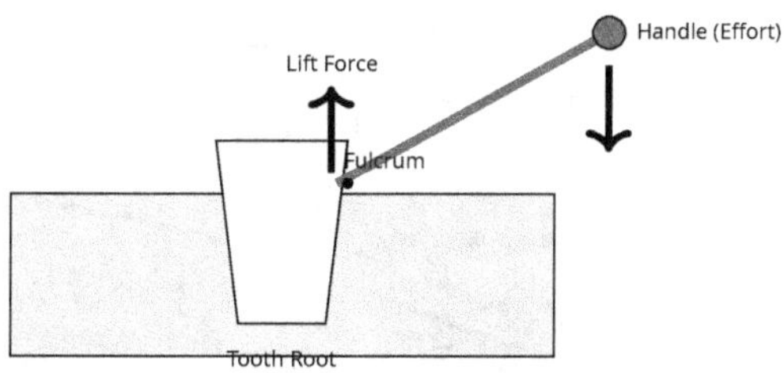

Figure 2: The dental elevator utilizes leverage to detach the tooth from its socket.

Biofilm and PH Dynamics

Oral diseases result from dental plaque, a bacterial biofilm that converts dietary sugars into acids. These acids demineralize the enamel structure.

Hygiene Components:

- **Abrasives:** Ground calcium carbonate (chalk) physically removes the biofilm.
- **Alkaline Rinses:** Sodium bicarbonate solutions neutralize the acidic plaque environment.
- **Surfactants:** Dilute soap facilitated debris removal through lower surface tension.

Regular removal of plaque prevents the formation of calculus (tartar), a mineralized deposit that requires mechanical removal.

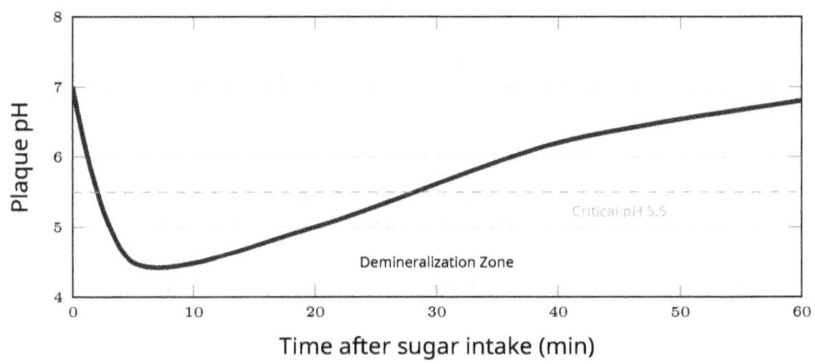

Figure 3: Enamel dissolution occurs when the oral pH level falls below the critical threshold.

Mechanical Calculus Removal

Calculus provides a porous surface that harbors pathogenic bacteria, leading to gingivitis and bone loss. Removal of these mineralized deposits requires mechanical scaling.

Scalers are sharp metal instruments designed to scrape calculus from the tooth surface and from below the gum line. Precise sharpening of dental instruments is required for effective debridement with minimal trauma to the surrounding tissues.

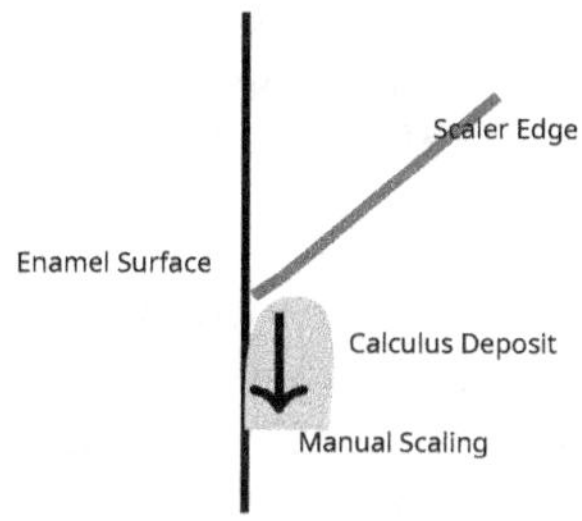

Figure 4: Mechanical scaling removes the bacterial reservoirs that cause periodontal disease.

Summary

Emergency dentistry manages septic pressure and maintains a neutral oral environment. Combined with hygiene and scaling, these techniques eliminate a primary source of chronic pain and systemic infection risk.

End of Discovery #040

DISCOVERY #041: PAPER

Category: Materials / Knowledge **Status:** Essential

The Molecular Felt

Paper is a thin material produced through the mechanical and chemical processing of cellulose fibers. These fibers, obtained from wood, rags, or hemp, are suspended in water and then deposited onto a screen to form a mat. As the water drains, the fibers interlock through hydrogen bonding to create a stable and flexible sheet.

Structural integrity depends on the length and purity of the cellulose chains. Mechanical pulping involves grinding raw materials to separate the fibers. Chemical pulping uses alkaline solutions, such as lye, to dissolve lignin. Lignin is the organic polymer that binds plant fibers. Its removal prevents the paper from becoming brittle and yellowing.

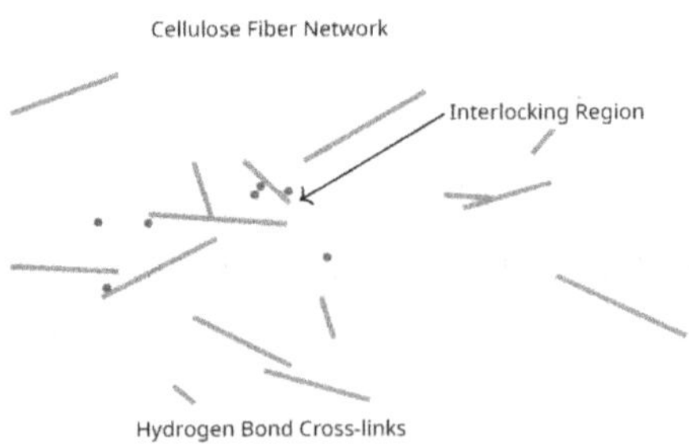

Figure 1: Random orientation and fiber entanglement provide multi-directional strength.

The Mold and Deckle

Sheet formation uses a **mold**, which consists of a fine wire mesh supported by a wooden frame. A second frame, the **deckle**, sits on top to define the sheet edges. The assembly is dipped into a vat containing dilute pulp.

Lifting the mold vertically allows water to drain through the mesh. This leaves a uniform layer of fibers on the surface. Shaking the mold during drainage ensures fibers distribute evenly. This agitation prevent the formation of a grain that would make the paper prone to tearing in a single direction.

Hydraulic Drainage

Gravity and surface tension govern the rate of water removal. High fiber-to-water ratios produce thick sheets for packaging. Lower ratios produce fine writing paper.

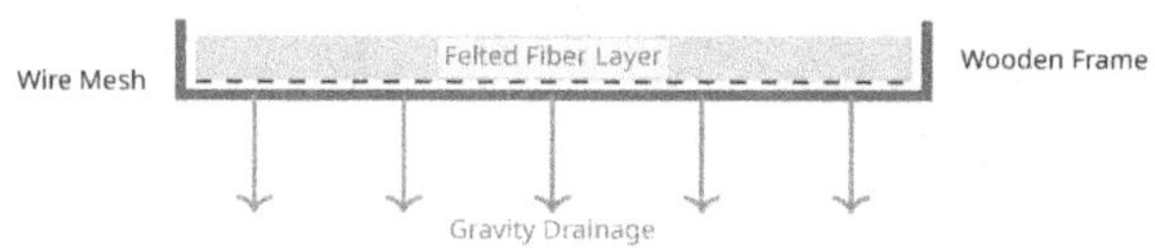

Figure 2: The mold acts as a selective filter for the liquid suspension.

Couching and Water Removal

Transferring the wet sheet from the mold onto absorbent felt is called **couching**. Multiple sheets and felts are stacked to form a post. A heavy screw press then removes the remaining bulk water and compacts the fibers.

Surface Sizing

Unfinished paper, or waterleaf, is highly absorbent. This absorption causes ink to spread. Suitability for writing or printing requires **sizing**. Sheets are dipped in a solution of gelatin or starch. The sizing agent fills microscopic voids between fibers. This process creates a smooth, non-absorbent surface that holds crisp ink lines.

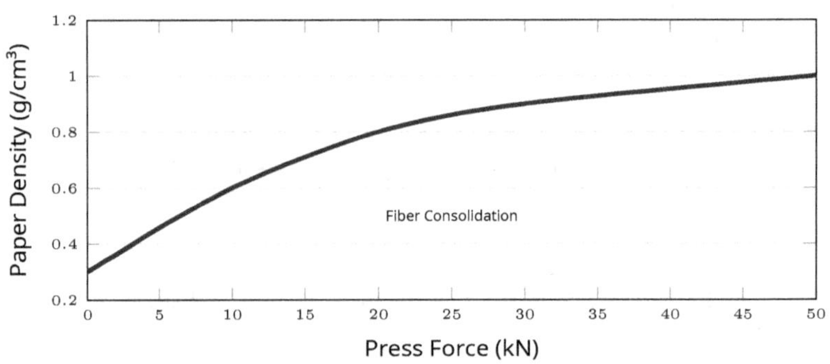

Figure 3: Mechanical pressure increases fiber density and improves surface smoothness.

The Archive of Civilization

Mass production of paper enables systematic storage of information. Durable and inexpensive paper supports the expansion of administration, commerce, and science.

Functional Properties:

- **Opacity:** Permits printing on both sides of the sheet.
- **Tensile Strength:** Enables mechanical handling in printing presses.
- **Longevity:** Alkaline-buffered paper persists for centuries in dry storage.

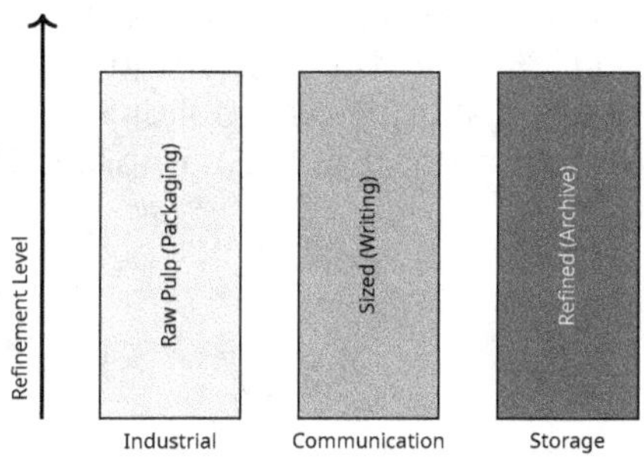

Figure 4: Industrial utility scales with the degree of fiber purification.

Summary

Paper is the universal medium for data transmission. Converting raw plant fibers into a standardized support for text establishes the infrastructure for collective memory.

End of Discovery #041

DISCOVERY #042: MOVABLE TYPE

Category: Mechanics / Communication **Status:** Essential

Modular Information

Movable type is a printing system that uses individual components for each character. These components are arranged in a frame to create a page of text. Mass-reproduction of text requires the creation of thousands of identical, durable characters with precise dimensions.

The system uses a specific **type-metal alloy**. A successful alloy consists of Lead (80%), Antimony (15%), and Tin (5%). Antimony provides hardness. It ensures the metal expands slightly as it solidifies. This expansion forces the metal into every detail of the mold. Tin improves fluidity and prevents oxidation.

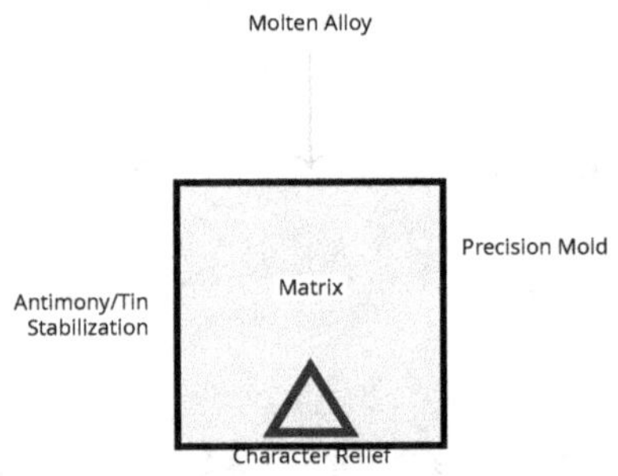

Figure 1: The hand-mold creates standardized characters with microscopic detail.

Oil-Based Ink

Traditional water-based inks are unsuitable for metal type because they bead on metallic surfaces. The printing press requires an **oil-based ink** with high viscosity.

The vehicle is linseed oil. Boiling the oil increases its thickness and drying speed. The primary pigment is **lampblack**, which is fine carbon soot. This combination produces a tacky paste that adheres evenly to metal type. It transfers cleanly to paper under pressure without blurring.

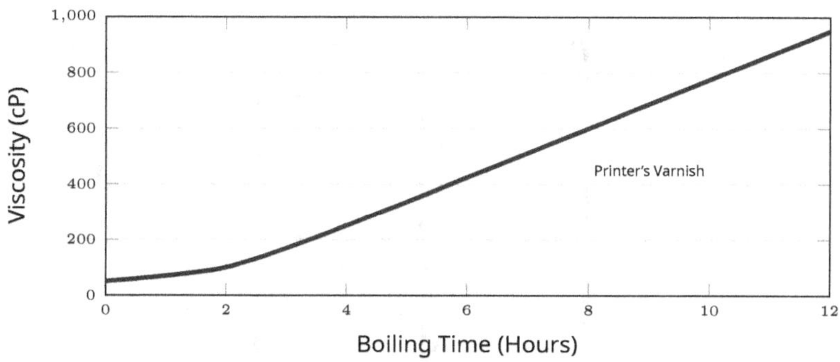

Figure 2: Controlled oxidation of linseed oil achieves the required printing viscosity.

Mechanical Pressure Distribution

The printing press converts rotational motion into high-pressure vertical force. The central component is a heavy wooden or iron screw called the **spindle**. Turning the handle drives the **platen** downward onto the paper and inked type.

Uniform pressure ensures a clean impression. The type is held in a rigid frame on a sliding bed. The operator inks the type, positions the paper, and slides the assembly under the platen for compression. Mechanical advantage replaces the inconsistent pressure associated with hand-rubbing.

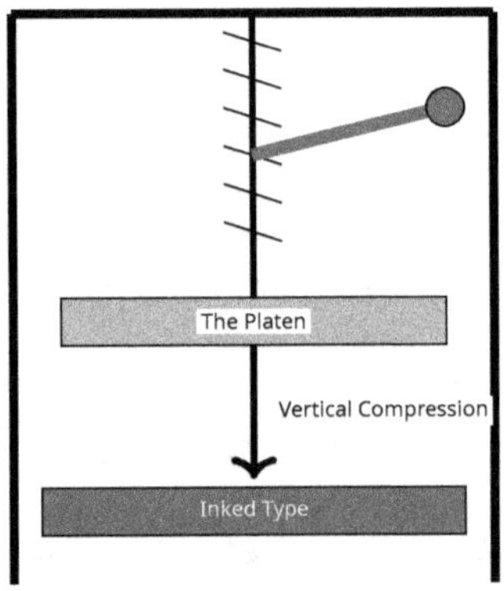

Figure 3: The screw mechanism provides the force necessary for clean ink transfer.

Scalability of Information

The printing press reduces the cost per unit of information. Initial **composition** requires significant labor. Once the type is set, hundreds of identical copies can be produced with minimal incremental effort.

This scalability supports the standardization of language and the synchronization of scientific observation. Technical diagrams and data tables can be distributed without the introduction of copying errors. Knowledge remains stable across time and space.

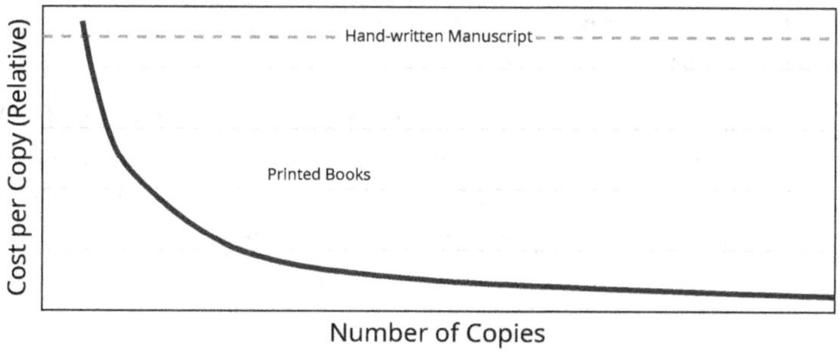

Figure 4: Automation reduces unit costs as volume increases.

Summary

Movable type establishes the mechanical foundation for the information age. By decoupling the reproduction of text from biological limits, a civilization ensures that its discoveries are preserved and shared.

End of Discovery #042

DISCOVERY #043: THE COMPASS

Category: Physics / Navigation **Status:** Essential

The Terrestrial Magnetic Field

Earth functions as a giant magnetic dipole. Convection currents within the molten iron core generate a planetary-scale magnetic field. This field consists of flux lines that emerge from the magnetic South Pole and converge at the magnetic North Pole.

A magnetized needle, when permitted to rotate freely on a low-friction pivot, aligns itself parallel to these flux lines. This alignment provides a consistent directional reference independent of celestial visibility. This discovery establishes the foundation for transoceanic navigation and precise land surveying.

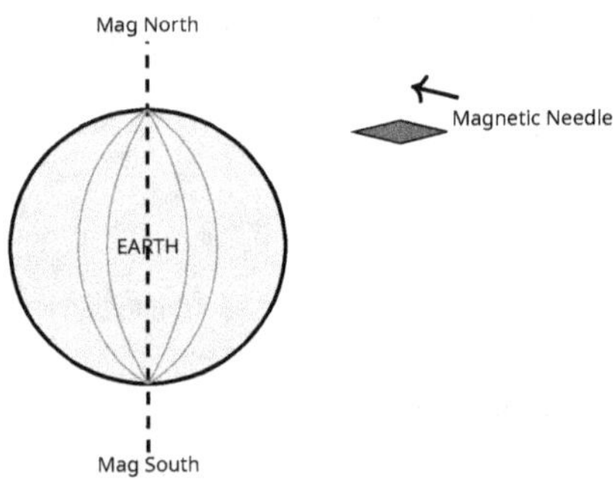

Figure 1: The planetary dipole forces magnetic materials into a North-South orientation.

Induced Magnetism

Magnetization involves the alignment of internal magnetic domains within a ferrous material. Natural **Lodestone** (Magnetite, Fe_3O_4) possesses permanent magnetism. Rubbing a steel needle against a lodestone transfers this orientation to the steel through magnetic induction.

Steel needles are the standard for navigation due to high carbon content and magnetic retentivity. The needle retains its orientation indefinitely when kept away from intense heat or mechanical shock. These conditions disrupt the atomic alignment of the domains.

Polarity Identification

The end of the needle pointing toward the North is designated the North-seeking pole. Identifying this pole is required for the correct calibration of navigation charts and course plotting.

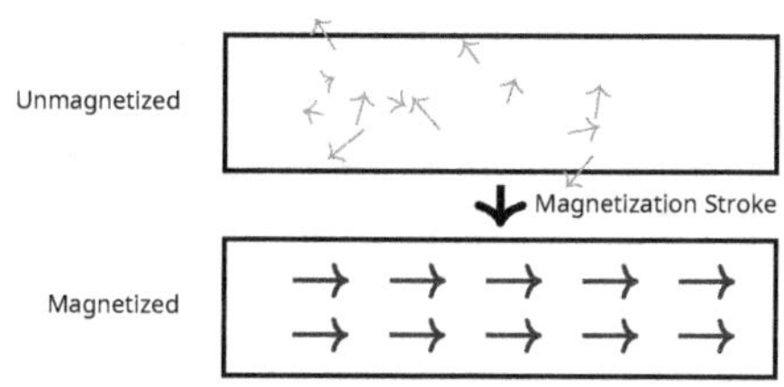

Figure 2: Alignment of internal magnetic domains converts steel into a directional tool.

Low-Friction Suspension

Compass utility depends on the sensitivity of the pivot mechanism. A mariner's compass utilizes a **gimbal system** to maintain a horizontal orientation. This system compensates for the rolling and pitching motions of a vessel.

The needle is attached to a circular card called the **compass rose**. This assembly rests on a hardened steel or jewel pivot. High-precision instruments utilize a liquid-filled chamber to provide damping. This fluid slows the oscillation of the card and allows for steady readings during turbulent sea conditions.

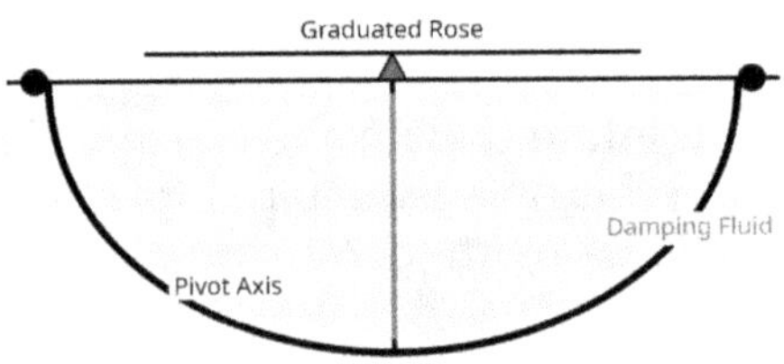

Figure 3: The gimbal and pivot assembly isolate the needle from mechanical interference.

The Variance of North

Magnetic North is distinct from True North. The angular difference between these two points is **Magnetic Declination**. This value varies according to geographic location and changes over time as the Earth's core shifts.

Accurate navigation requires local declination corrections to the compass reading. Failure to account for this variance results in course errors over long distances. Navigation charts include **Isogonic lines** to allow navigators to synchronize their instruments with geographic reality.

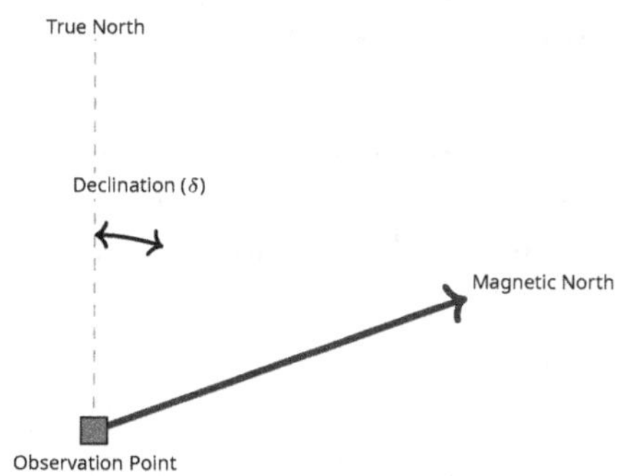

Figure 4: Correction for magnetic declination is required for precise course plotting.

Summary

The compass provides a reliable directional constant. Harnessing the magnetic properties of Earth enables the mapping of continents and the safe transit of the global oceans.

End of Discovery #043

DISCOVERY #044: MECHANICAL CLOCK

Category: Mechanics / Chronometry **Status:** Essential

The Regulation of Energy

A mechanical clock converts the continuous potential energy of a falling weight or a coiled spring into discrete, equal intervals. The primary component of this system is the **escapement**.

The escapement regulates energy release by allowing the gear train to advance by a single tooth at each oscillation of a controller. This mechanism produces the audible ticks of the clock. Accuracy depends on the isochronism of the oscillator, where each oscillation duration remains constant regardless of the amplitude.

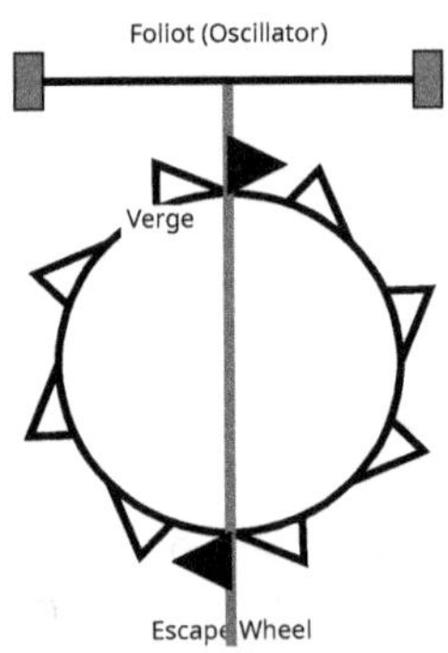

Figure 1: The escapement permits the gear train to advance in precise increments.

Rotational Ratios

The clock uses gears to translate the rapid movement of the escapement into the slower movements of the hands. This translation is achieved through specific **gear ratios**.

The power source drives the main wheel. This force is transmitted through the center wheel, the third wheel, and finally the escape wheel. The number of teeth on each gear ensures that the minute hand completes one rotation for every 3,600 oscillations of the oscillator. Precision tooth geometry is required to prevent jamming.

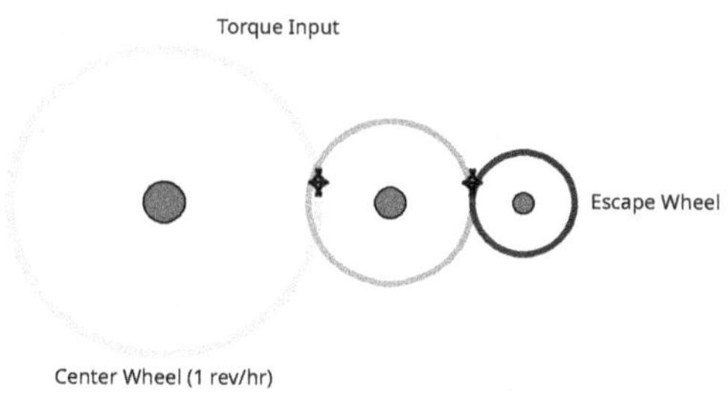

Figure 2: Multiple gear stages reduce speed while increasing the duration of operation.

Isochronism and Temperature

Mechanical clocks are sensitive to environmental factors. Thermal expansion causes pendulums to lengthen and balance wheels to expand, slowing the clock. High-precision chronometry requires **thermal compensation**.

The gridiron pendulum uses rods of steel and brass with opposing expansion coefficients. As steel expands downward, brass expands upward by an equal amount. This keeps the effective center of mass constant. Bimetallic balance wheels curve inward during heating to compensate for the weakening of the balance spring.

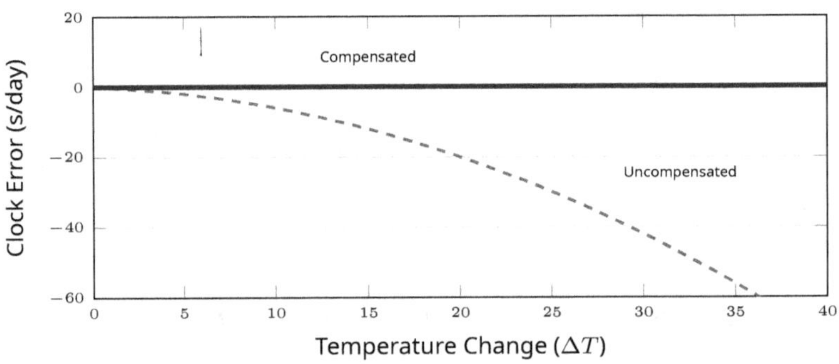

Figure 3: Compensation mechanisms neutralize the effects of thermal expansion.

Navigation and Time

Determining longitude at sea requires knowing the time difference between the local position and a reference meridian. Earth rotates 15 degrees per hour. One hour of time difference equals 15 degrees of longitude.

The **Marine Chronometer** is a clock designed for extreme precision in variable environments. By maintaining reference time, a navigator can compare it to local noon to calculate their exact East-West position. Local noon is determined by the highest point of the sun.

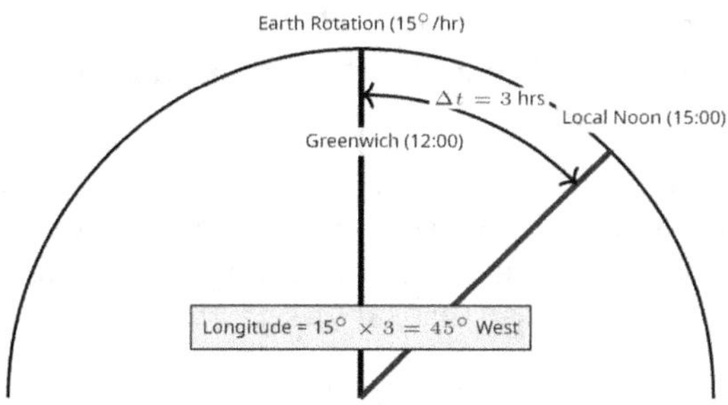

Figure 4: Longitude is derived from the time difference between locations.

Summary

The mechanical clock transforms time into a measurable spatial coordinate. Achieving micro-second precision in variable environments enables the mapping of the planet with absolute accuracy.

End of Discovery #044

DISCOVERY #045: OPTICAL LENSES

Category: Physics / Optics **Status:** Essential

The Science of Refraction

Optical lenses utilize the refractive properties of glass to manipulate light paths. Refraction occurs as light transitions between media of different optical densities, such as air and glass. The speed of light decreases within the glass medium, causing the wavefront to bend according to the surface curvature.

A **Convex Lens** (converging) is thicker at the center. It directs parallel rays toward a single focal point. This concentration enables real image formation on a surface. A **Concave Lens** (diverging) is thinner at the center. It spreads rays outward, creating virtual images.

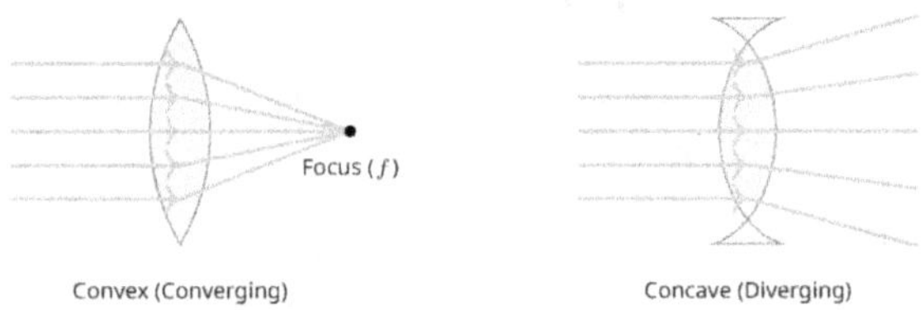

Figure 1: Geometric surface curvature determines the convergence or divergence of light rays.

Correction of Optical Errors

Human vision depends on focusing light precisely onto the retinal plane. Variations in eyeball shape produce refractive errors.

Myopia (Nearsightedness): Light focuses in front of the retina. A concave lens diverges incoming rays to shift the focal point backward. **Presbyopia (Farsightedness):** Light focuses behind the retina. A convex lens converges rays to move the focal point forward.

Spectacles restore visual acuity for technical observation. Standardized lens powers, measured in diopters, allow for the industrial production of corrective optics.

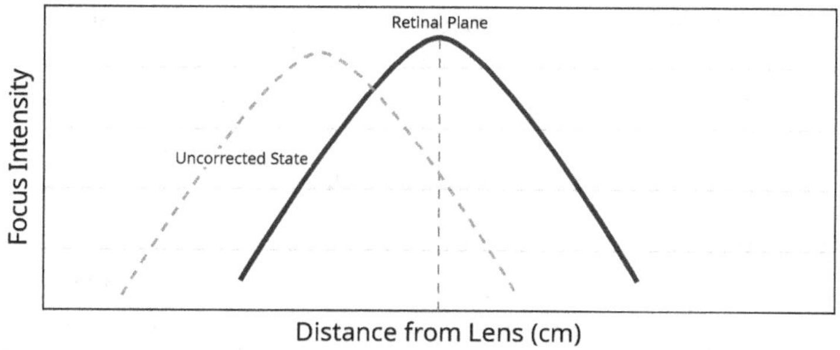

Figure 2: Lenses reposition the focal peak to compensate for anatomical deviations.

Optical Amplification

The telescope is an instrument designed to collect a volume of light greater than the human eye and to magnify distant objects. The refracting telescope utilizes two primary lenses in series within a light-shielded tube.

The **Objective Lens** possesses a long focal length. It collects light and creates a primary image at its focal plane. The **Eyepiece Lens** possesses a short focal length. It acts as a magnifying glass to examine the primary image.

Total magnification is the ratio of focal lengths:

$$\text{Magnification} = \frac{f_{\text{objective}}}{f_{\text{eyepiece}}}$$

This discovery enables precise navigation and astronomical observation.

Figure 3: The objective lens concentrates light for magnification by the ocular lens.

Grinding and Polishing

Manufacturing precision optics requires material removal through controlled abrasion. A glass blank is rubbed against a metal tool, or **lap**, of the required curvature.

Process Sequence:

- **Generating:** Rough grinding using emery to create the basic curve.
- **Smoothing:** Using fine grits to eliminate surface pits.
- **Polishing:** Applying iron oxide on a soft pitch bed.

Surface deviations must be smaller than the wavelength of light. This precision prevents light scattering and ensures image clarity in industrial and scientific instruments.

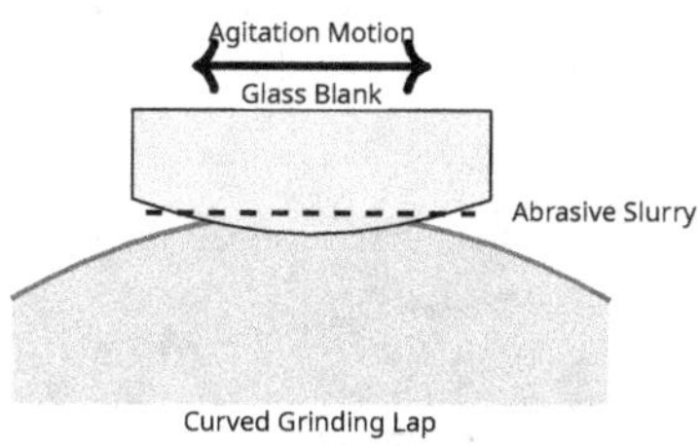

Figure 4: Reciprocating motion over a curved lap generates spherical lens surfaces.

DISCOVERY #046: THE TELEGRAPH

Category: Electricity / Communication **Status:** Essential

The Solenoid Principle

The telegraph utilizes electromagnetism to transmit signals over copper conductors. Electric current passing through a solenoid generates a magnetic field. This field attracts a ferrous armature, moving a lever mechanism.

This mechanical movement converts electrical pulses into physical motion at a distant location. Solenoid activation occurs as the circuit closes. An operator sends discrete signals through the conductive medium by controlling the current duration.

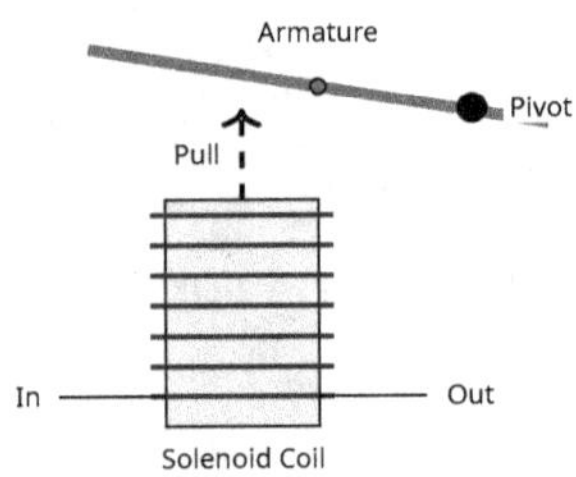

Figure 1: Electrical current generates the force required to move a distant receiver.

Pulse Modulation

Communication requires a standardized protocol for translating pulses into language. This is achieved through time-based modulation of the signal.

Binary Pulse Coding (Morse):

- **Dot:** A short duration pulse.
- **Dash:** A pulse three times the duration of a dot.
- **Spacing:** Silence intervals between letters and words.

This modulation enables the transmission of the entire alphabet using a single wire. Operators identify characters by the rhythmic pattern of electromagnetic clicks.

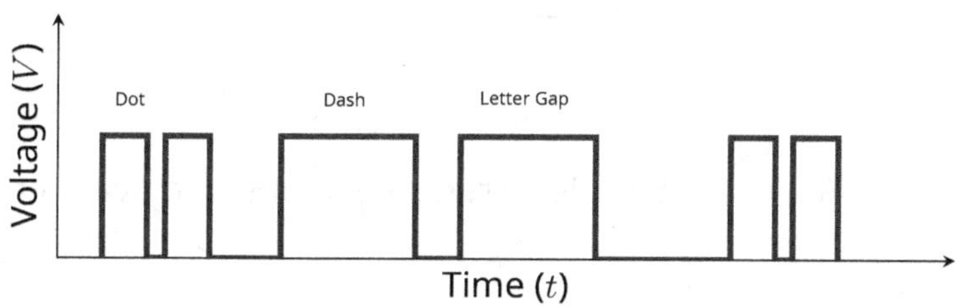

Figure 2: Pulse duration determines the information content of the transmission.

The Conductive Loop

Telegraphy requires a complete circuit between transmitter and receiver. A suspended wire carries the signal. Ground return systems minimize infrastructure costs.

The earth acts as the return conductor in a ground return system. This requires metal plates buried in moist soil at both ends. Batteries provide the potential difference required to overcome the electrical resistance of long-distance wires.

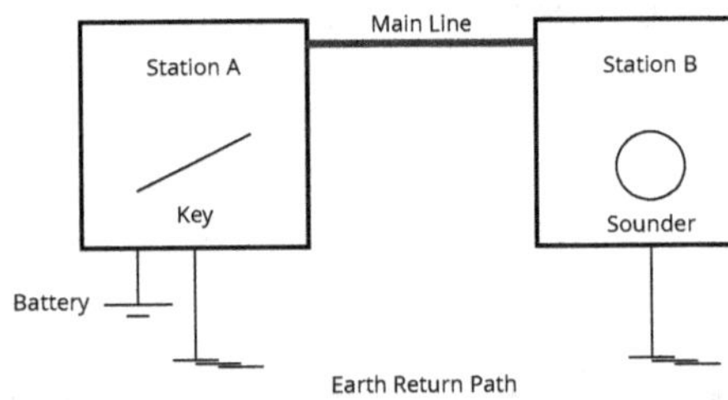

Figure 3: The earth provides a return path for electrical signals.

Signal Regeneration

Voltage decreases as a signal travels through a wire due to resistance. The current eventually becomes too weak to activate a sounder. The **Relay** solves this limitation.

A relay is a sensitive electromagnet that acts as an automated switch. A weak incoming signal activates the relay, which closes a local circuit powered by its own battery. This local circuit transmits a high-voltage signal to the next line section. Relays enable inter-continental network creation.

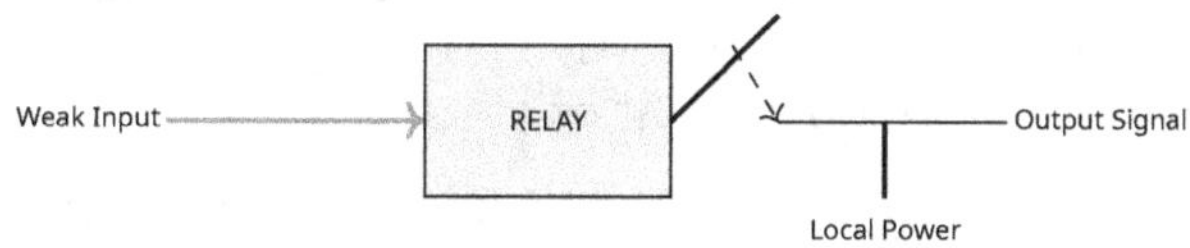

Figure 4: Relays act as signal amplifiers by triggering local power sources.

Summary

The telegraph establishes the communication infrastructure of an industrial society. Decoupling information from physical transit enables centralized management across planetary distances.

End of Discovery #046

DISCOVERY #047: BASIC ELECTRICITY

Category: Physics / Energy | **Status:** Essential

Chemical Potential Energy

The Voltaic Pile is the primary instrument for generating a continuous flow of electricity. It operates on the principle of electrochemical potential. When two dissimilar metals are separated by an electrolyte, a chemical reaction occurs that facilitates the movement of electrons.

A standard pile consists of alternating disks of zinc and copper. These disks are separated by cardboard soaked in a brine or acid solution. Zinc acts as the negative electrode, or anode, and releases electrons through oxidation. Copper acts as the positive electrode, or cathode. Stacking these units in series increases the total electromotive force of the system.

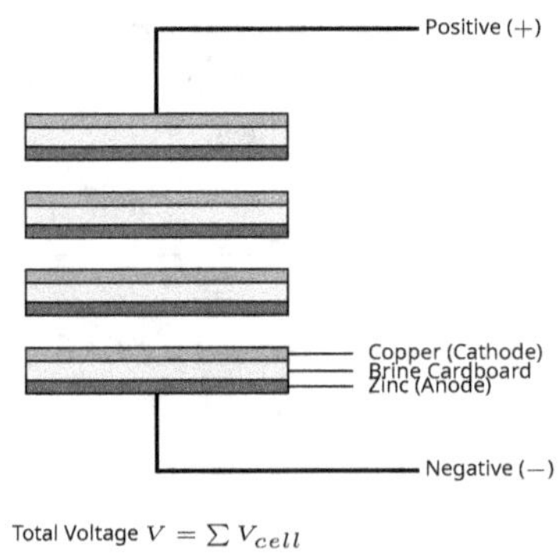

Figure 1: Series stacking of chemical cells multiplies the output voltage.

The Faraday Principle

The relationship between magnetism and electricity enables the conversion of mechanical work into electrical energy. Electromagnetic induction occurs when a conductor moves through a changing magnetic field.

As a magnetic flux passes through a loop of wire, it induces an electrical current. The magnitude of this induced current is proportional to the rate of change of the magnetic flux. This discovery allows for the generation of large-scale power through the rotation of coils near magnets.

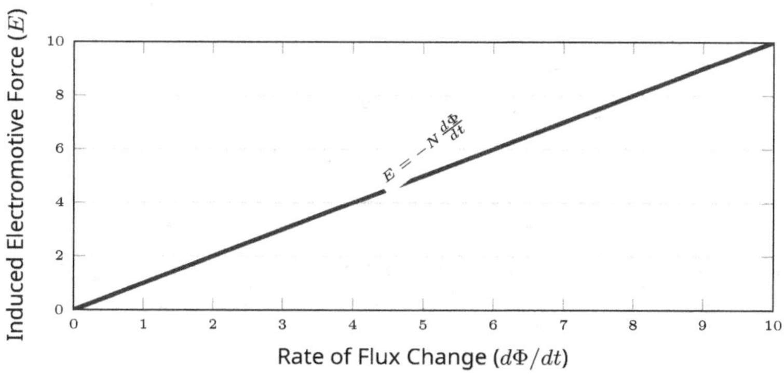

Figure 2: Induced voltage increases linearly with the rate of magnetic flux change.

Mechanical Power Conversion

The dynamo is a mechanical generator that produces direct current through electromagnetic induction. It consists of a rotating assembly of wire coils, known as the armature, positioned within a stationary magnetic field.

A commutator ensures a constant direction of current. As the armature rotates, the direction of the induced current in the coils reverses periodically. The commutator acts as a mechanical rotary switch, reversing the connection to the external circuit in synchronization with the rotation. This resulting unidirectional current is suitable for charging batteries and powering telegraph systems.

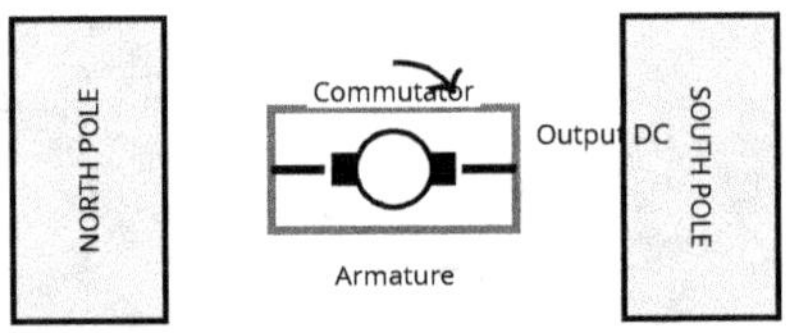

Figure 3: The commutator translates alternating internal current into direct external current.

Electrical Flow Resistance

Electrical circuit behavior is governed by the relationship between potential, flow, and resistance. Voltage acts as the pressure, while current represents the volume of flow.

Resistance is a physical property of the conductor. Longer and thinner wires increase resistance, resulting in energy dissipation as heat. Efficient power transport requires conductors with a cross-section sufficient to minimize these losses.

Ohmic Power Law: The rate of energy transfer, measured in Watts (P), is the product of voltage (V) and current (I).

$$P = V \cdot I = I^2 \cdot R$$

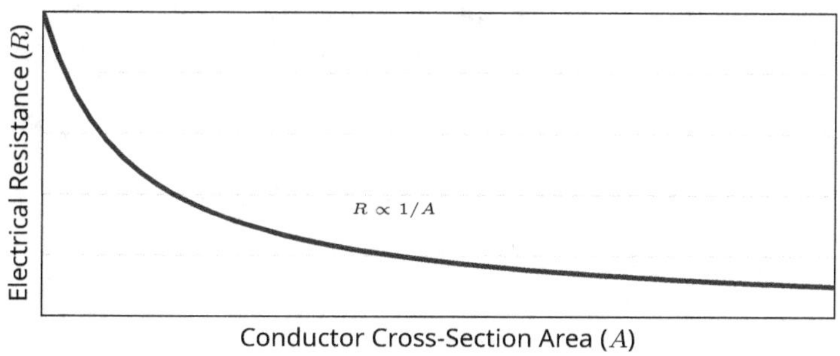

Figure 4: Increasing the conductor area reduces resistive energy dissipation.

Summary

Basic electricity enables the efficient transmission of energy. Utilizing chemical and mechanical sources establishes a scalable power medium for industrial processes and global telecommunications.

End of Discovery #047

DISCOVERY #048: SPARK-GAP RADIO

Category: Communication / Physics **Status:** Essential

Hertzian Wave Propagation

Radio communication relies on the propagation of electromagnetic waves through space. These waves are generated by the acceleration of electrical charges. In a spark-gap system, an electrical discharge across an air gap creates a rapidly oscillating high-frequency current.

This oscillation generates expanding spheres of electric and magnetic fields. These waves travel at the speed of light and induce a corresponding current in distant conductors. This physical phenomenon enables the transmission of information without physical wire connections.

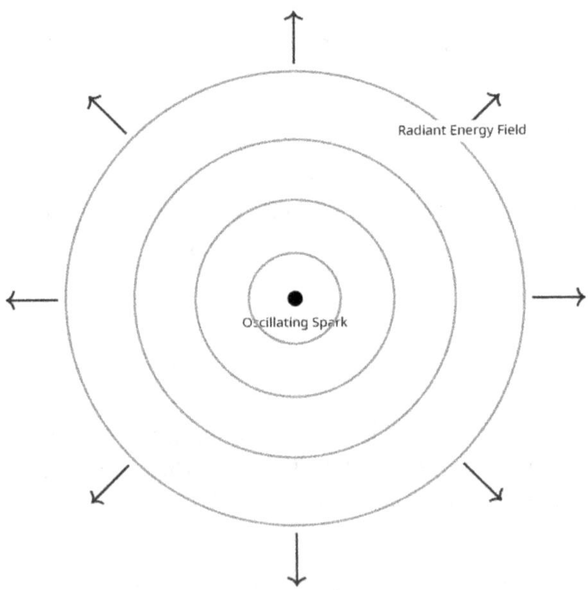

Figure 1: Accelerated electrical charges radiate energy as spherical electromagnetic waves.

The Induction Coil

Radio wave generation requires a high-voltage source. The induction coil utilizes electromagnetic induction to transform low-voltage direct current into thousands of volts. A mechanical interrupter rapidly breaks the primary circuit, inducing a voltage spike in the secondary winding.

The Spark Gap Discharge

This high voltage connects to two metal electrodes separated by an air gap. When the voltage exceeds the dielectric strength of air, a spark occurs. This discharge is a high-frequency oscillation between the electrodes and the antenna system. Antenna dimensions determine the fundamental frequency of the emitted wave.

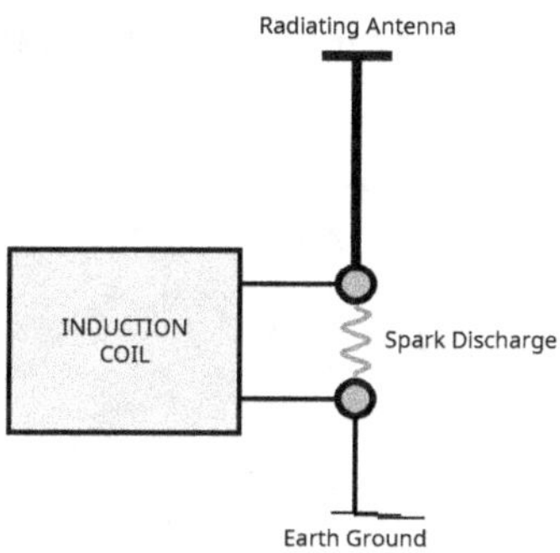

Figure 2: The antenna system radiates high-frequency energy from the spark gap.

Detection of Radio Waves

Detecting faint radio signals requires a sensitive, non-linear device. The coherer is the primary receiver for spark-gap systems. It consists of a glass tube filled with loose metal filings, typically nickel and silver, positioned between two electrodes.

In a resting state, the filings possess high electrical resistance. When a radio wave strikes the antenna, the oscillating fields cause the filings to cohere, drastically reducing their resistance. This transition allows a local battery to trigger a sounder. A mechanical device shakes the tube after each signal to reset the filings for the next pulse.

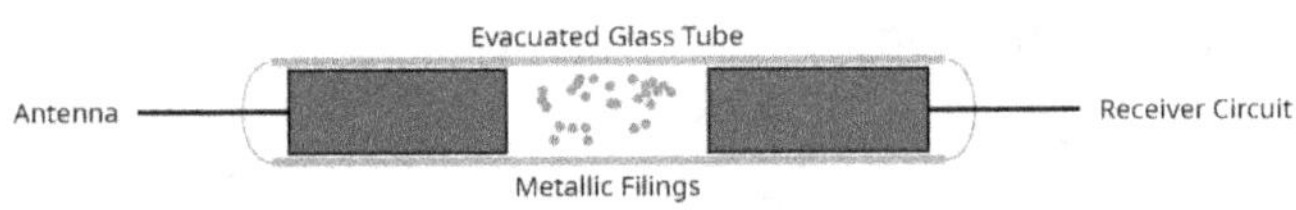

Figure 3: Radio waves act as a switch by altering the conductivity of metal particles.

Frequency Isolation

Un-tuned transmitters emit broadband signals that occupy a wide range of frequencies, causing interference. Adding a capacitor and an inductor to both the transmitter and receiver creates a resonant circuit.

The station operates at a specific frequency defined by the physical properties of its components. Resonance permits the receiver to respond only to waves of a specific length. This isolation significantly increases the effective communication range and allows multiple stations to operate simultaneously.

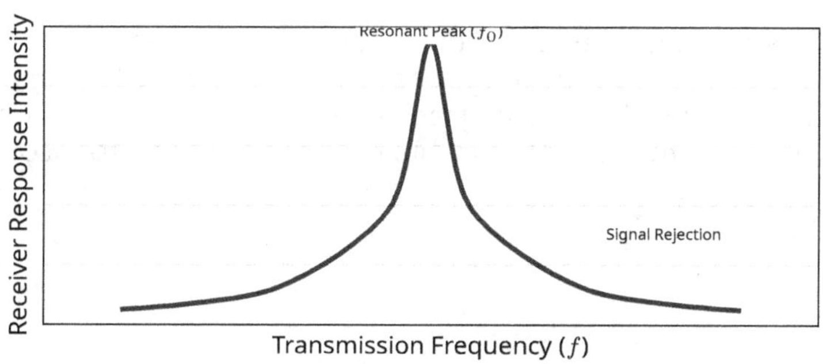

Figure 4: Resonant circuits filter the desired signal from background electromagnetic noise.

Summary

Spark-gap radio is the first practical application of electromagnetic radiation. By manipulating wave propagation and resonance, a civilization overcomes physical barriers, enabling wireless data transmission across planetary distances.

End of Discovery #048

DISCOVERY #049: MUSICAL INSTRUMENTS

Category: Acoustics / Physics **Status:** Essential

The Physics of Sound

Sound is a longitudinal wave produced by the mechanical vibration of a medium. Musical instruments control these vibrations to produce specific frequencies known as pitches.

A vibrating string generates standing waves. The fundamental frequency is determined by the string's length, tension, and mass. Halving the length of a vibrating string doubles its frequency, producing a pitch one octave higher. This mathematical relationship provides the basis for acoustic engineering and the construction of scaled instruments.

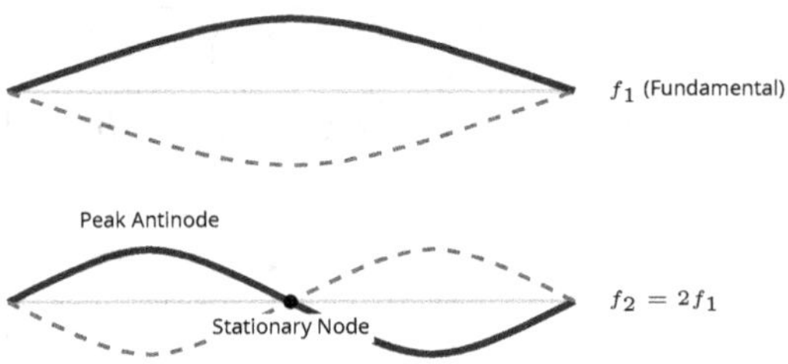

Figure 1: Physical constraints on a string define the harmonic progression.

The Chordophone Mechanism

The lute is a chordophone that utilizes a resonance chamber to amplify string vibrations. The strings are stretched over a bridge, which transfers vibrational energy to a thin wooden soundboard.

The hollow body acts as a Helmholtz resonator. Air inside the chamber vibrates in sympathy with the soundboard, increasing the volume and richness of the tone. Specific soundboard thickness and cavity volume ensure efficient sound wave projection. This mechanical amplification allows the instrument to produce audible levels suitable for group environments.

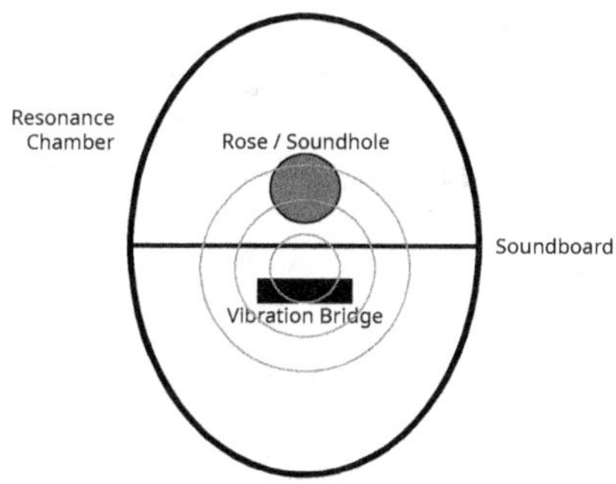

Figure 2: The instrument's internal volume provides mechanical amplification.

The Aerophone Principle

The flute is an aerophone that produces sound through the oscillation of an air column. Air directed against a sharp labium edge creates pressure fluctuations.

The length of the tube determines the wavelength of the internal standing wave. Finger holes modify the effective length of the air column to select resonant frequencies. A tube open at both ends produces a fundamental frequency where the wavelength is twice the tube length.

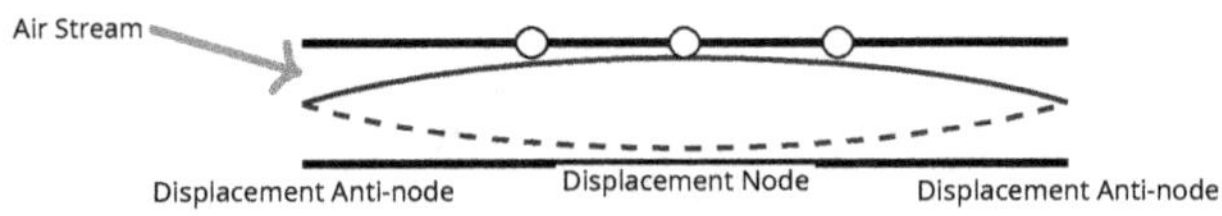

Figure 3: Finger holes modify the effective length of the resonant air column.

Mathematical Harmony

Geometric progressions define the organization of pitch scales. Small integer ratios produce stable intervals. These ratios align the overtones of notes to prevent acoustic interference.

- **Octave (2:1):** Total consonant stability.
- **Perfect Fifth (3:2):** High harmonic stability.
- **Perfect Fourth (4:3):** Structural harmonic stability.

Standardizing these intervals allows for the collective synchronization of multiple instruments. Melodic patterns derived from these ratios facilitate the preservation of complex information.

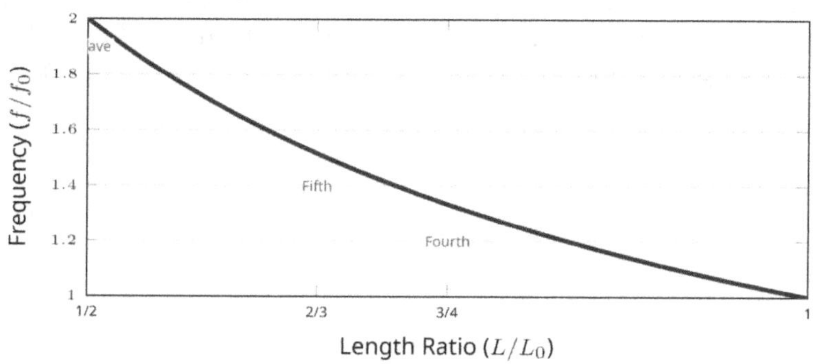

Figure 4: The inverse relationship between length and frequency defines the scale.

Summary

Musical instruments translate physics into a standardized sound language. Mastering the acoustics of vibrating systems secures a medium for cultural transmission and the mathematical analysis of nature.

End of Discovery #049

DISCOVERY #050: THE SOCIAL CONTRACT

Category: Social Science / Law **Status:** Essential

The Basis of Civil Order

The Social Contract is a society's implicit agreement to cooperate for mutual benefit. A structured legal framework enables security and resource management. This transition supports long-term planning and the systematic protection of property.

The stability of the contract relies on the consent of the governed and the equitable enforcement of established rules. Surrendering absolute individual autonomy to a central authority secures collective safety. This order allows for the protection of individual rights through institutional mechanisms.

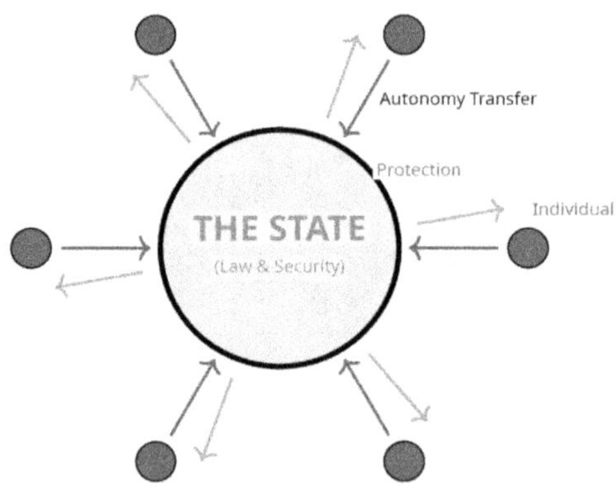

Figure 1: Institutional security is maintained through the delegation of autonomy.

Codified Systems

A legal system provides the architectural framework for society. Codified rules apply universally to ensure predictability in social and economic interactions. This universality prevents arbitrary governance.

Core Pillars:

1. **Legality:** Acts are punishable solely through existing laws.
2. **Due Process:** Individuals have the right to transparent adjudication.
3. **Enforcement:** The state holds the exclusive authority to uphold the law through force.

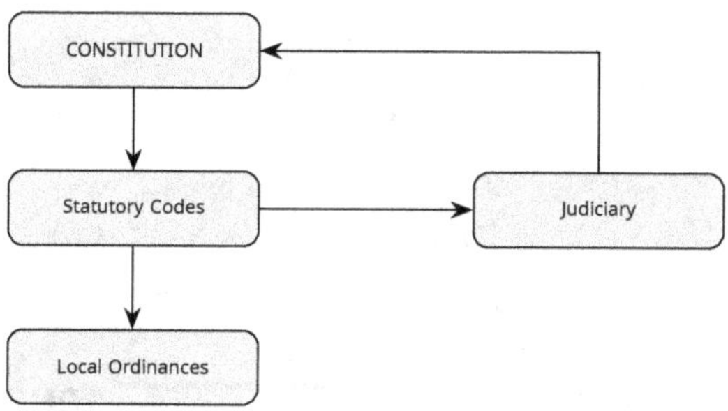

Figure 2: Hierarchy ensures consistent application of law across the population.

The Medium of Exchange

Economic expansion requires a standardized medium for value transfer. Currency provides a universal proxy for labor and resources. This medium overcomes the limitations of direct commodity trade.

A stable currency performs three primary functions:

- **Unit of Account:** Provides a common measure for goods.
- **Medium of Exchange:** Facilitates immediate transactions.
- **Store of Value:** Preserves economic power across time.

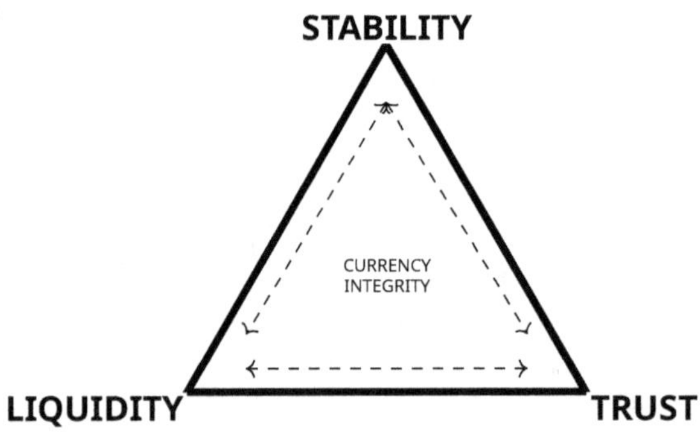

Figure 3: Economic stability relies on the equilibrium of trust and liquidity.

Minting and Ledgers

Trust in currency originates from physical and administrative standardization. The state mints coins with fixed metallic content and weight. The official stamp certifies value and eliminates the requirement for repeated measurement.

Economic Infrastructure:

- **The Mint:** A secure facility for producing standardized currency units.
- **The Ledger:** A central record of credits and obligations.
- **Taxation:** A mechanism for resource collection to maintain infrastructure and defense.

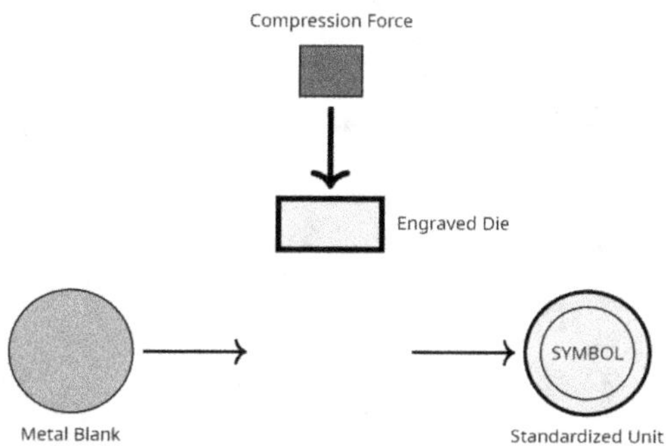

Figure 4: Minting creates interchangeable units of account for commerce.

Summary

The Social Contract and economic systems decouple survival from physical strength. Establishing universal laws and standardized exchange creates the cooperative environment necessary for industrial and scientific progress.

End of Discovery #050

www.ingramcontent.com/pod-product-compliance
Lightning Source LLC
LaVergne TN
LVHW052324200926
848481LV00002B/658
9781806475230